# ATLAS OF THE
# HUMAN BODY

# ATLAS OF THE

# HUMAN BODY

## ANATOMY · PHYSIOLOGY · HEALTH

BARRON'S

First edition for the United States, its territories and dependencies, and
Canada published in 2008 by Barron's Educational Series, Inc.

© Copyright of English language translation 2008 by Barron's Educational
Series, Inc.

Original title of the book in Spanish: *El Gran Libro Del Cuerpo Humano*
Copyright © 2006 by Parramón Ediciones, S.A.—World Rights
Published by Parramón Ediciones, S.A., Barcelona, Spain

Author: Parramón's Editorial Team
Illustrations: Parramón's Archive, Marcel Socías
Photographs: AGE Fotostock, Parramón's Archive, Getty Images, Manel
Clemente, Prisma
Text: Adolfo Cassan

English translation by Gerald L. Geiger

*All inquiries should be addressed to:*
Barron's Educational Series, Inc.
250 Wireless Boulevard
Hauppauge, NY 11788
www.barronseduc.com

ISBN-13: 978-0-7641-6091-2
ISBN-10: 0-7641-6091-5

Library of Congress Control Number: 2007924987

Printed in China
9 8 7 6 5 4 3 2 1

# Foreword

**The Big Book of the Human Body** is a reference work on anatomy, physiology, and health. All members of the family, from teenager to senior citizen, should find this work helpful in understanding how the body works as well as how to take care of it.

Using colorful illustrations of body parts and organs, the book is divided into the various organ systems of the body. The illustrations are accompanied by information that describes the basic anatomy and physiology of the area being studied. An exhaustive index easily enables readers to locate answers to their questions.

It is our hope that our readers will increase their understanding of medical science by becoming familiar with the topics presented in this book.

# Contents

# Introduction

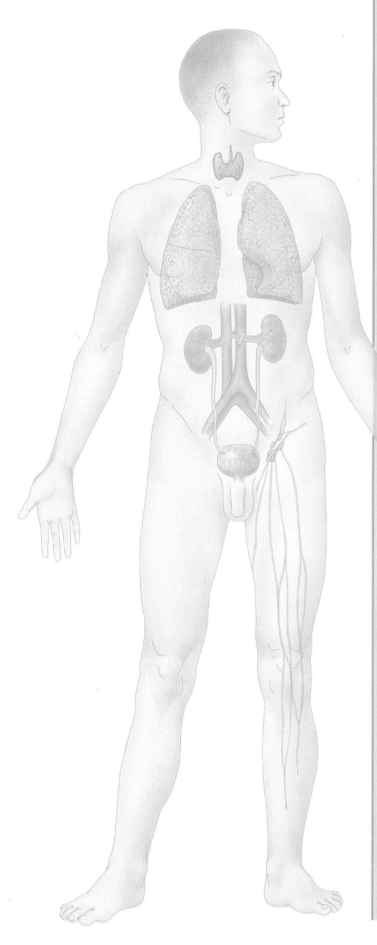

It is a truism that the human body is infinitely more complex than the most complicated machine. It is simply a self-evident statement because nothing could resemble the wonder of the body. In contrast to any artificial device, the body is shaped, grows, and is regenerated independently. It maintains a constant exchange with the external environment from which it gets what it needs for survival. The human body ceaselessly adapts to the constant changes that take place around it. To continue living and thriving, it has millions and millions of components that are uniquely interrelated.

**Cells and tissues**

The human body is made up of an enormous number of cells, which are the basic units of any living thing. It is estimated that the body of an adult contains more than 200 billion cells. Although these cells all contain the same basic components, they feature different shapes and are equipped to perform distinct specific functions. These differentiated cells are not arranged in a random pattern. Depending on their characteristics, the cells are grouped and, on occasion, are combined with inert substances, such as minerals, salts, or fibers produced by the cells, to form tissues.

Basically four types of tissues are in the human body, each with special missions. First is the epithelial tissue, whose most important functions are those related to covering and secretion. Second is the connective tissue, made up of a variety of cell types. For instance, protein fibers provide support to the body structures. Third is the muscle tissue, made up of elongated cells that are capable of contracting in response to a stimulus and then resuming their initial dimensions in order to provide mobility for the body. Fourth is nerve tissue, which consists of cells capable of both receiving and generating stimuli in the form of electric impulses. These transmit information that governs the activity of the muscles and the endocrine glands or that governs and initiates thought processes.

The specialization of some cells and the combination of different tissues give rise to the formation of specific organs responsible for performing specific tasks, such as the skin, the stomach, the liver, the lungs, and the heart.

### Organs

Each organ has a particular location and a specific function: some are solid (liver, spleen) and others are hollow (stomach, intestines). Some organs have tissues that are not present in any other part of the body, such as the epidermis, which is the surface layer of the skin, or bone tissue, which is the principal component of bones. On the other hand, there are widely different organs whose properties depend on the presence of the same tissue type. The numerous muscles of the body, the heart, and various hollow viscera, for example, can contract and expand because they have muscle tissue. What characterizes the organs is their specific function.

### Organ systems

Organs perform particular functions. The skin, which provides protection for all internal structures, has other functions as well. There are some organs that carry out their functions in combination with other organs (liver, kidneys).

Many systems are made up of different organs that are composed of various types of tissues. For example, the digestive system is made up of

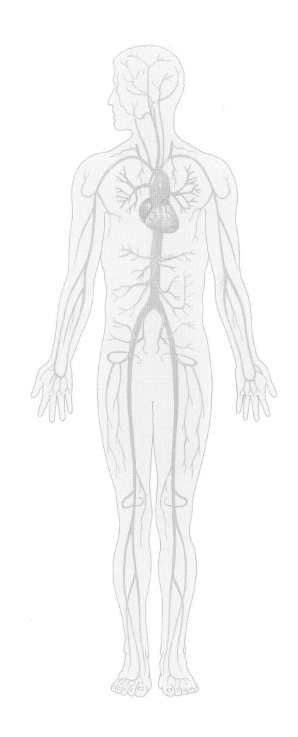

structures such as the mouth, the stomach, and the liver. The respiratory system is made up of the nose, the larynx, the bronchi, and the lungs. The circulatory system is made up of the heart, the arteries, and the veins.

Bone and muscle tissue together form the musculoskeletal system. The skeletal system and the muscular system essentially are made up, respectively, of bone tissue and muscle tissue. The endocrine system consists of different glands that secrete hormones into the blood.

All of the systems, however, are interrelated. The functions of each can be performed fully only by interacting with the others. All are necessary to make up a fully functional body. In the organs and systems mentioned earlier, the digestive system is responsible for nutrition and the respiratory system enables us to obtain oxygen from the environment. The circulatory system makes it possible for blood, which carries nutrients and oxygen to all the tissues, to move around the body. The locomotor system enables us to perform the movements that are required for daily life. There are many other organs just as important as these—for example, the sense organs and the organs of the urogenital system.

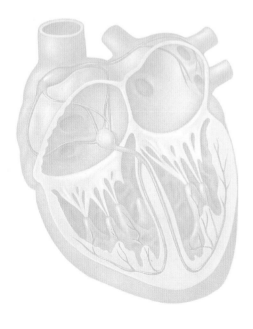

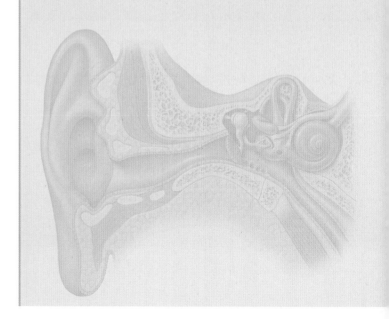

Throughout the book, we will describe both the components of the cells and the different parts of the human body. Then we will present the principal concepts of the anatomy and the physiology of the various systems of the human body.

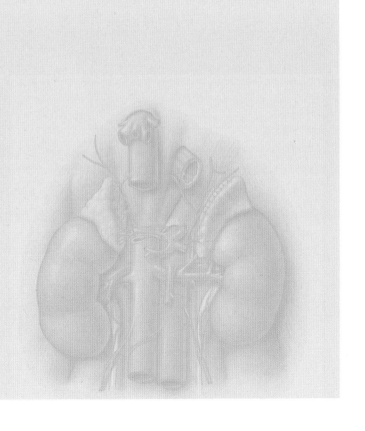

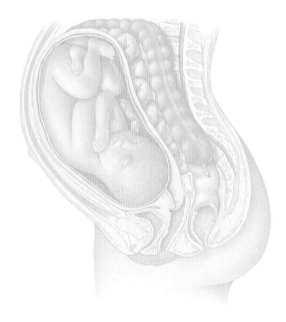

# Cells and chromosomes

organelles indispensable for carrying out the metabolic functions necessary for life and communicates with the external environment to maintain cellular integrity.

## Components of the human cell

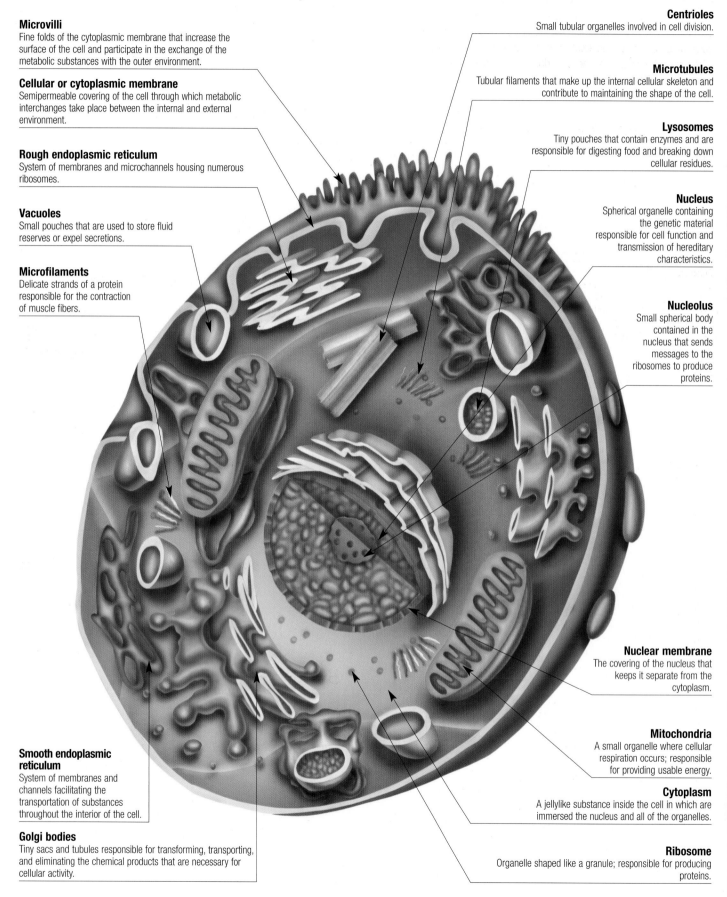

**Microvilli**
Fine folds of the cytoplasmic membrane that increase the surface of the cell and participate in the exchange of the metabolic substances with the outer environment.

**Cellular or cytoplasmic membrane**
Semipermeable covering of the cell through which metabolic interchanges take place between the internal and external environment.

**Rough endoplasmic reticulum**
System of membranes and microchannels housing numerous ribosomes.

**Vacuoles**
Small pouches that are used to store fluid reserves or expel secretions.

**Microfilaments**
Delicate strands of a protein responsible for the contraction of muscle fibers.

**Smooth endoplasmic reticulum**
System of membranes and channels facilitating the transportation of substances throughout the interior of the cell.

**Golgi bodies**
Tiny sacs and tubules responsible for transforming, transporting, and eliminating the chemical products that are necessary for cellular activity.

**Centrioles**
Small tubular organelles involved in cell division.

**Microtubules**
Tubular filaments that make up the internal cellular skeleton and contribute to maintaining the shape of the cell.

**Lysosomes**
Tiny pouches that contain enzymes and are responsible for digesting food and breaking down cellular residues.

**Nucleus**
Spherical organelle containing the genetic material responsible for cell function and transmission of hereditary characteristics.

**Nucleolus**
Small spherical body contained in the nucleus that sends messages to the ribosomes to produce proteins.

**Nuclear membrane**
The covering of the nucleus that keeps it separate from the cytoplasm.

**Mitochondria**
A small organelle where cellular respiration occurs; responsible for providing usable energy.

**Cytoplasm**
A jellylike substance inside the cell in which are immersed the nucleus and all of the organelles.

**Ribosome**
Organelle shaped like a granule; responsible for producing proteins.

## DNA, chromosomes, and genes

All of the information that governs the development and activity of an organism is stored in deoxyribonucleic acid (DNA). This DNA makes up the chromosomes that are present in the nucleus of the cells and the basic functional units—the genes.

DNA is made up of two long, parallel, microscopic chains intertwined with each other in the form of a double helix. DNA contains three basic ingredients: phosphate molecules, molecules of a five-carbon sugar named deoxyribose, and four types of nitrogenous bases: adenine, guanine, thymine, and cytosine. Each chain consists of thousands of basic units known as nucleotides; hydrogen bonds link two chains of nitrogenous bases so that the DNA double helix looks like a structure resembling a spiral staircase.

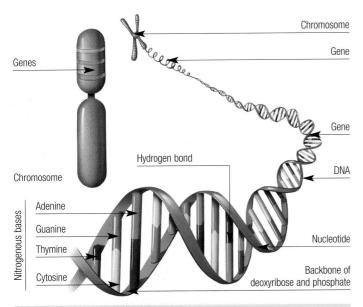

Genes

Chromosome

Nitrogenous bases

Adenine

Guanine

Thymine

Cytosine

Hydrogen bond

Chromosome

Gene

Gene

DNA

Nucleotide

Backbone of deoxyribose and phosphate

## Structure of chromosomes

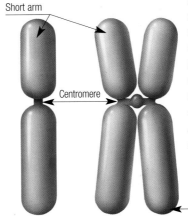

Short arm

Centromere

Long arm

Chromosome before replication of DNA

Chromosome after replication of DNA

*Although chromosomes vary in size, they all have a similar shape: a tiny cylinder with a constriction, the centromere, that divides the chromosome into two arms of differing lengths. This typical image of the chromosome resembles a stage of cell division in which DNA has already replicated, resembling an X with two short arms and two long arms.*

## Chromosomal makeup

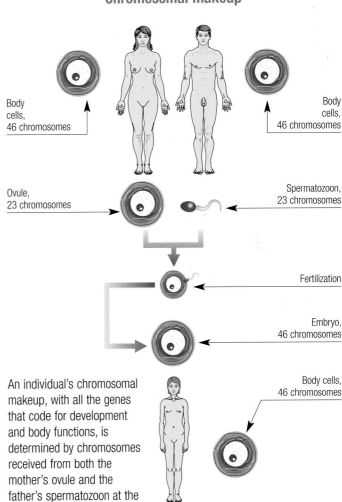

Body cells, 46 chromosomes

Body cells, 46 chromosomes

Ovule, 23 chromosomes

Spermatozoon, 23 chromosomes

Fertilization

Embryo, 46 chromosomes

Body cells, 46 chromosomes

An individual's chromosomal makeup, with all the genes that code for development and body functions, is determined by chromosomes received from both the mother's ovule and the father's spermatozoon at the moment of fertilization. This is possible because, in contrast to the other cells of the human body which have 46 chromosomes, the gametes have only 23 chromosomes. Specifically, the merging of one ovule and one spermatozoon will give rise to an egg cell with 46 chromosomes (23 pairs of homologous chromosomes) from whose successive divisions will spring an embryo made up of cells that have the identical chromosomal makeup. Thus, every individual receives one-half of his or her chromosomes from the mother and the other half from the father.

## Replication of DNA

Division of all cells of the body, except the germ cells, is preceded by the duplication of the chromosomal material because each daughter cell must receive an exact copy of DNA from the parent cell. In this process, called replication, the two DNA chains are separated. The actions of the enzyme DNA polymerase cause two new DNA chains to form, each identical to the parent molecule. This enzyme uses the backbones of the parent molecule as templates to create complementary chains. The strict bonding requirements of the nitrogenous bases (adenine bonding only with thymine and cytosine bonding only with guanine) result in the formation of these daughter chains. As a result, two identical molecules are formed, each containing one original DNA chain and one new chain.

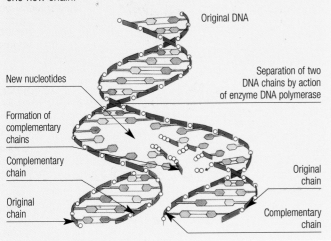

New nucleotides

Formation of complementary chains

Complementary chain

Original chain

Original DNA

Separation of two DNA chains by action of enzyme DNA polymerase

Original chain

Complementary chain

# Genes and heredity

## Genetic code

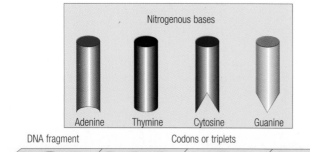

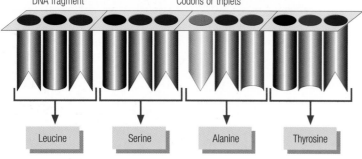

Coded amino acids

The genetic code is the information required for the synthesis of the proteins, which consist of a combination of amino acids. There are thousands of different proteins, but all are formed by the combination of twenty different amino acids. The genes code for the specific order of the individual amino acids needed to form a protein. Although this seems very complicated, the fact is that the mechanism that governs the genetic code is quite simple. It is based on the sequence of the nitrogenous bases of the DNA.

The four types of nitrogenous bases form a kind of alphabet that is read by looking at it in groups of three. Each triplet or codon codes for an amino acid. The order of triplets determines the makeup of the polypeptide chain. This genetic code is identical in all living species. It is, as a matter of fact, universal.

## Dominant and recessive genes

Every person receives half of his or her genetic makeup from each parent and, therefore, two units of each gene. Each gene, therefore, is present in each of the two homologous chromosomes and is situated in a specific place called a locus. Some genes, although they have the same function, come in variants called alleles. For example, the gene that determines eye color has variants that cause the iris to adopt a bluish tone or a brownish tone. Sometimes the information contained in an allele is imposed upon the one that is contained in the other. The first one is therefore called "dominant," while the second one is called "recessive."

Not all genes are necessarily expressed. Some are dominant. Only one copy of a dominant allele in a homologous pair is necessary for the dominant trait to be expressed. In contrast, recessive genes are expressed only when the recessive allele is present in both chromosomes of the pair. In the example, the allele for brown eye color is dominant. Therefore, only one copy of the allele is needed for the individual to have brown eyes. In contrast, the allele for blue eye color is recessive. An individual will have blue eyes only when both of the homologous chromosomes contain this recessive allele for blue eye color.

## Inheritance of eye color

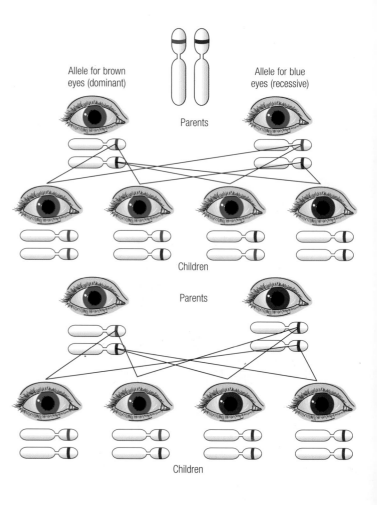

Allele for brown eyes (dominant)

Allele for blue eyes (recessive)

Parents

Children

Parents

Children

All of the cells have 46 chromosomes with the exception of the gametes, the ova and the spermatozoa, which have only half. Because of their identical shape and size, homologous chromosomes form pairs. Human cells therefore contain 23 pairs of chromosomes. In 22 of these pairs, called autosomes, both chromosomes are homologous. The one remaining pair, the sex chromosomes, may or may not be homologous. In females, both sex chromosomes are X chromosomes. They are identical in shape and size. The two X chromosomes are homologous. In males, the sex chromosomes consist of one X chromosome and one Y chromosome. They differ greatly in size and shape. They are not homologous. The transmission of anatomical and physiological characteristics from parents to offspring, including the transmission of certain diseases, is governed by the location of the specific genes and whether the alleles are dominant or recessive.

### Types of autosomal inheritance

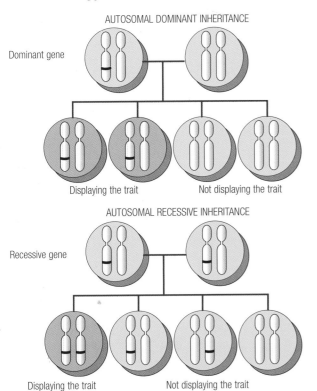

AUTOSOMAL DOMINANT INHERITANCE

Dominant gene

Displaying the trait    Not displaying the trait

AUTOSOMAL RECESSIVE INHERITANCE

Recessive gene

Displaying the trait    Not displaying the trait

### Types of heredity tied to the X chromosome

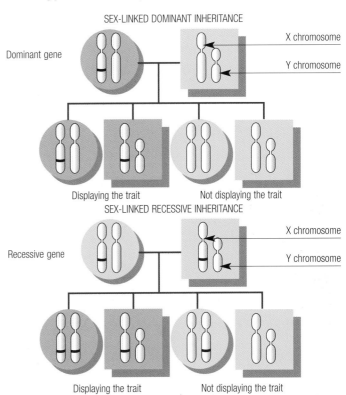

SEX-LINKED DOMINANT INHERITANCE

Dominant gene

X chromosome

Y chromosome

Displaying the trait    Not displaying the trait

SEX-LINKED RECESSIVE INHERITANCE

Recessive gene

X chromosome

Y chromosome

Displaying the trait    Not displaying the trait

*We speak of autosomal dominant inheritance when a trait or a disease depends on the presence of a dominant gene in an autosomal chromosome. The dominant gene comes from only one parent. The recessive gene is eclipsed by the dominant one. On the other hand, we speak of autosomal recessive inheritance when a trait or a disease depends on the presence of two recessive genes in the chromosome pair. This gene must be transmitted by both the father and the mother.*

*Heredity tied to the X chromosome can be either dominant or recessive in the case of the female, but it is always dominant in the male because the latter has a single X chromosome.*

### Heredity tied to the Y chromosome

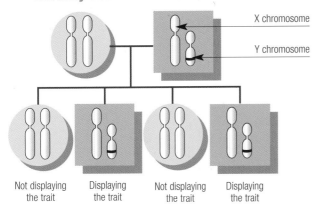

X chromosome

Y chromosome

Not displaying the trait    Displaying the trait    Not displaying the trait    Displaying the trait

*In heredity tied to the Y chromosome, the feature or the disease are always and only manifested in the male since this sex chromosome is absent in the chromosomal makeup of the female.*

### The human genome

The term genome is applied to the complete set of genes of a species. As a result of a study undertaken by a large group of investigators, it is now possible to decipher the genome of the human individual by analyzing approximately 3.5 million pairs of the nitrogenous bases contained in the 46 chromosomes. The study made it possible to identify that only about 35,000 genes are responsible for coding proteins, just a small portion of the chromosomal DNA. The remainder, by virtue of mechanisms that so far are not well-known, act to modulate the entire process. Because of this, in each cell, some genes code for proteins while others remain inactive. Different cell types express different genes, although all cells have the same genetic makeup. How specific genes "know" to code for proteins in certain cells and not in others is still a mystery. We already know a lot about the human genome—and still have much more to learn.

# Parts of the body

**The human body is made up of diverse organ systems, each one adapted for a** specific function. All of them interrelated so that the activity of the whole, which enables us to maintain a suitable measure of health and to enjoy life, will develop in a coordinated fashion.

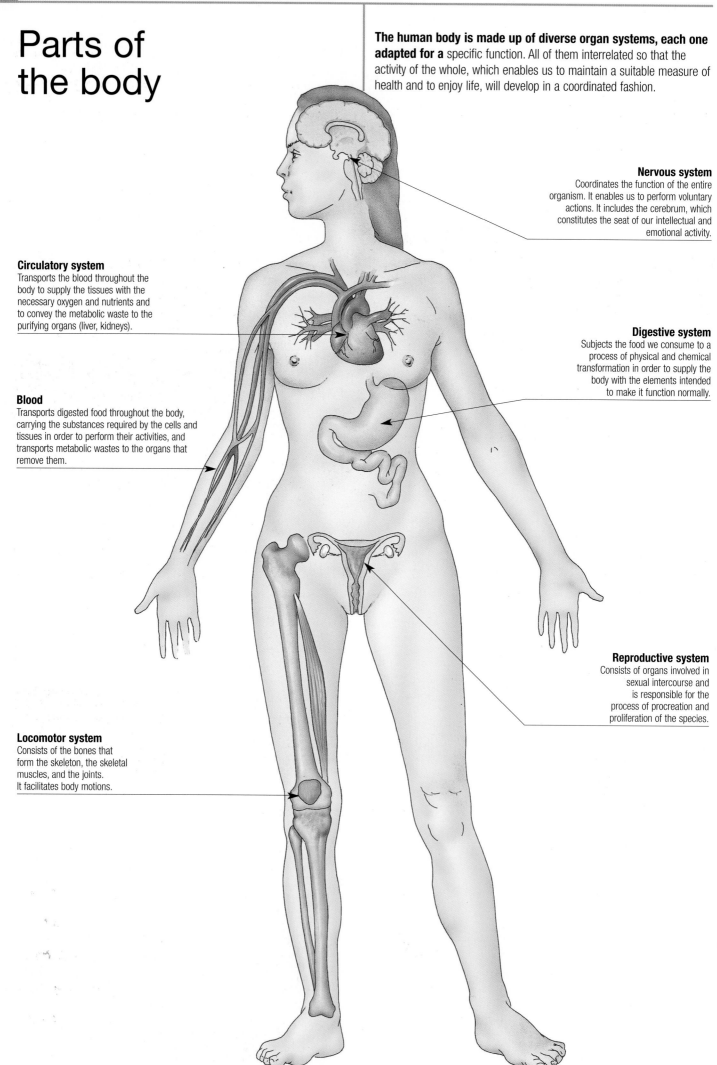

**Nervous system**
Coordinates the function of the entire organism. It enables us to perform voluntary actions. It includes the cerebrum, which constitutes the seat of our intellectual and emotional activity.

**Circulatory system**
Transports the blood throughout the body to supply the tissues with the necessary oxygen and nutrients and to convey the metabolic waste to the purifying organs (liver, kidneys).

**Digestive system**
Subjects the food we consume to a process of physical and chemical transformation in order to supply the body with the elements intended to make it function normally.

**Blood**
Transports digested food throughout the body, carrying the substances required by the cells and tissues in order to perform their activities, and transports metabolic wastes to the organs that remove them.

**Reproductive system**
Consists of organs involved in sexual intercourse and is responsible for the process of procreation and proliferation of the species.

**Locomotor system**
Consists of the bones that form the skeleton, the skeletal muscles, and the joints. It facilitates body motions.

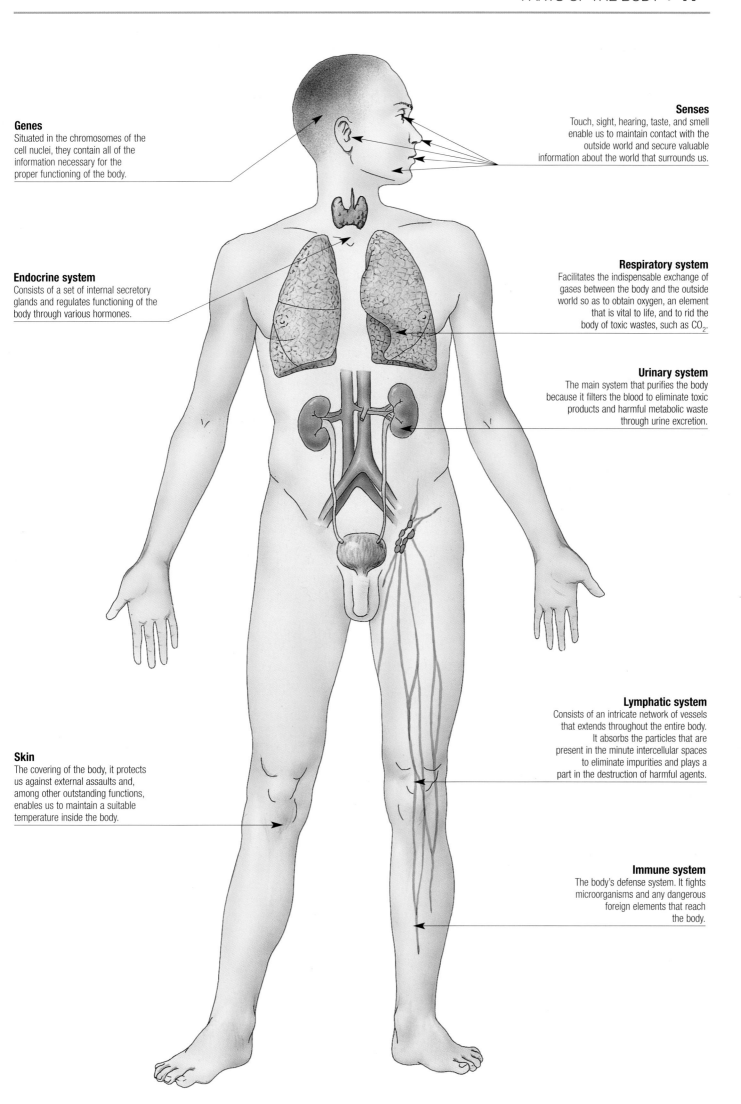

**Genes**
Situated in the chromosomes of the cell nuclei, they contain all of the information necessary for the proper functioning of the body.

**Endocrine system**
Consists of a set of internal secretory glands and regulates functioning of the body through various hormones.

**Skin**
The covering of the body, it protects us against external assaults and, among other outstanding functions, enables us to maintain a suitable temperature inside the body.

**Senses**
Touch, sight, hearing, taste, and smell enable us to maintain contact with the outside world and secure valuable information about the world that surrounds us.

**Respiratory system**
Facilitates the indispensable exchange of gases between the body and the outside world so as to obtain oxygen, an element that is vital to life, and to rid the body of toxic wastes, such as $CO_2$.

**Urinary system**
The main system that purifies the body because it filters the blood to eliminate toxic products and harmful metabolic waste through urine excretion.

**Lymphatic system**
Consists of an intricate network of vessels that extends throughout the entire body. It absorbs the particles that are present in the minute intercellular spaces to eliminate impurities and plays a part in the destruction of harmful agents.

**Immune system**
The body's defense system. It fights microorganisms and any dangerous foreign elements that reach the body.

# The bones

**The bones are hard and resistant parts of the body and come in** different shapes and sizes. They make up the framework of the body, providing protection for the vital organs and facilitating body motions. They are essential components of the locomotor system.

## Bone tissue

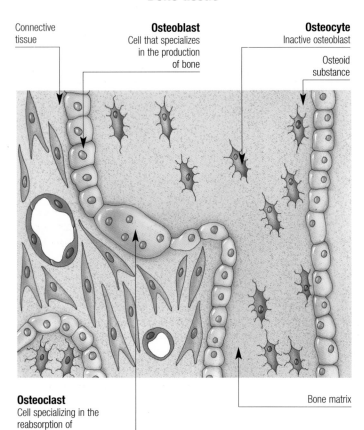

Connective tissue

**Osteoblast**
Cell that specializes in the production of bone

**Osteocyte**
Inactive osteoblast

Osteoid substance

**Osteoclast**
Cell specializing in the reabsorption of devitalized bone tissue

Bone matrix

*Bone tissue is made up of a matrix that consists of cells, collagenous fibers, and an amorphous substance on which calcium and phosphorus are deposited. These minerals give hardness to the bones. This tissue has some specialized cells that, due to the influence of various hormones, reshape the bones over time. Some reabsorb the devitalized tissue, and others develop new material.*

### Function of bones

- They are the framework of the body. They determine its shape and size.
- Linked to the muscles and tendons, which use the bones as levers to perform movements, they enable us to maintain posture, to walk, and to perform gestures.
- They protect certain parts of the body—above all, the soft and vulnerable organs.
- They constitute an important organic reserve of minerals, such as calcium and phosphorus.
- They contain the bone marrow that produces the blood cells.

## Three-dimensional drawing illustrating bone structure

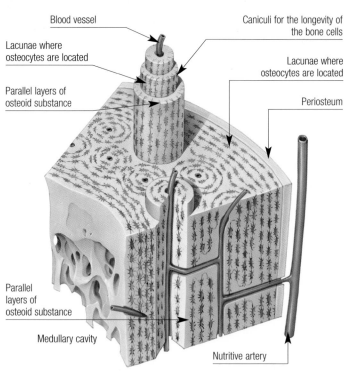

Blood vessel

Caniculi for the longevity of the bone cells

Lacunae where osteocytes are located

Lacunae where osteocytes are located

Parallel layers of osteoid substance

Periosteum

Parallel layers of osteoid substance

Medullary cavity

Nutritive artery

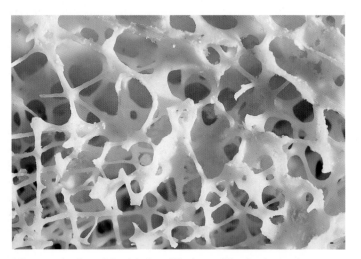

*Microscopic view of the interior of the bone. We clearly see the trabecular features of the bone tissue that form arches resembling those of a cathedral.*

*The osteoid substance is arranged in a special way. In the outer part of the bone, surrounded by a layer of resistant tissue called the periosteum, the bone layers are concentric. They are arranged around a central conduit through which a blood vessel can pass. They are traversed by multiple caniculi that provide passage for the contents of the above conduit. The entire complex, where the bone layers are tightly touching against each other without any cracks in between, forms a hard mass that provides strength for the bone and is called the compact bone tissue. On the other hand, inside the bone, the bone layers are arranged in irregular trabeculae that leave space between each other, thus constituting a spongy bone tissue that is less dense and has a porous appearance.*

## Types of bones

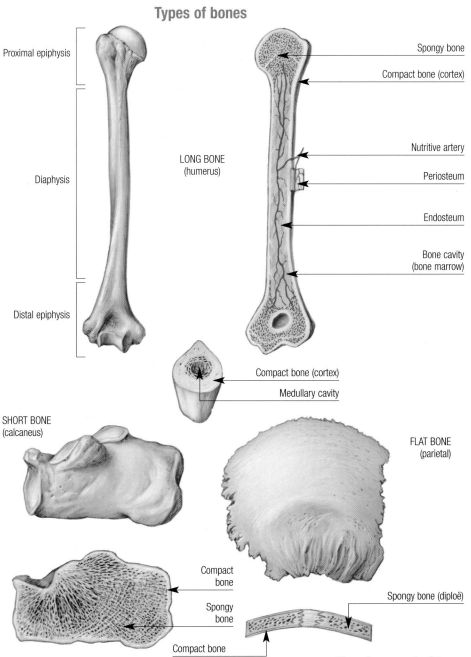

Proximal epiphysis

Diaphysis

Distal epiphysis

LONG BONE
(humerus)

Spongy bone

Compact bone (cortex)

Nutritive artery

Periosteum

Endosteum

Bone cavity
(bone marrow)

Compact bone (cortex)

Medullary cavity

SHORT BONE
(calcaneus)

FLAT BONE
(parietal)

Compact
bone

Spongy
bone

Compact bone

Spongy bone (diploë)

Although all the bones are made up of bone tissue and each has a characteristic shape, there are three types:

• **Long bones:** They are cylindrical with an enlarged central body, the **diaphysis**, and two ends, the **epiphyses**, which join the adjacent bones. The epiphyses consist of an outer layer of compact bone tissue with a thickness of several millimeters—the **cortex**, which is covered on the outside by a hard membrane, the **periosteum**, and supported on the inside by a strong membrane, the **endosteum**. The ends contain a spongy bone tissue whose trabeculae house the red bone marrow, where red blood cells are made. In the diaphysis, the outer layer constitutes the boundary of a bone space, the marrow cavity, which is occupied by yellow bone marrow that is inactive.

• **Flat bones:** They have various shapes and dimensions, but they are more or less wide. They consist of two layers of compact bone tissue that constitute the boundary of a narrow space occupied by spongy bone. This is called diploic, in whose trabeculae it also shelters bone marrow.

• **Short bones:** They are small in size, cylindrical or cubic, although their form varies. They consist of a thin layer of compact bone and are full of spongy bone tissue that, on occasion, contains bone marrow.

## Development of bone mass according to sex and age

*Bone formation starts during pregnancy. Its formation takes a long time because the skeleton completes its development by the end of adolescence. Bone mass increases progressively during childhood and undergoes a noteworthy increase during adolescence. Starting with the third decade of life, it begins to diminish, although, under normal conditions, bones continue to be sufficiently strong well into advanced age.*

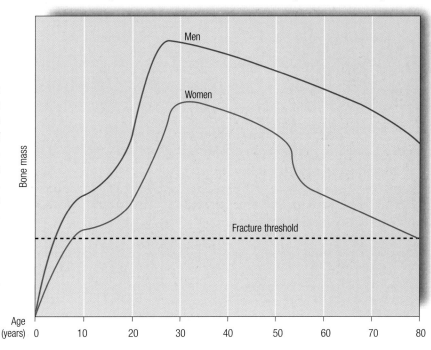

Bone mass

Men

Women

Fracture threshold

Age
(years)  0    10    20    30    40    50    60    70    80

# Bones: the skeleton

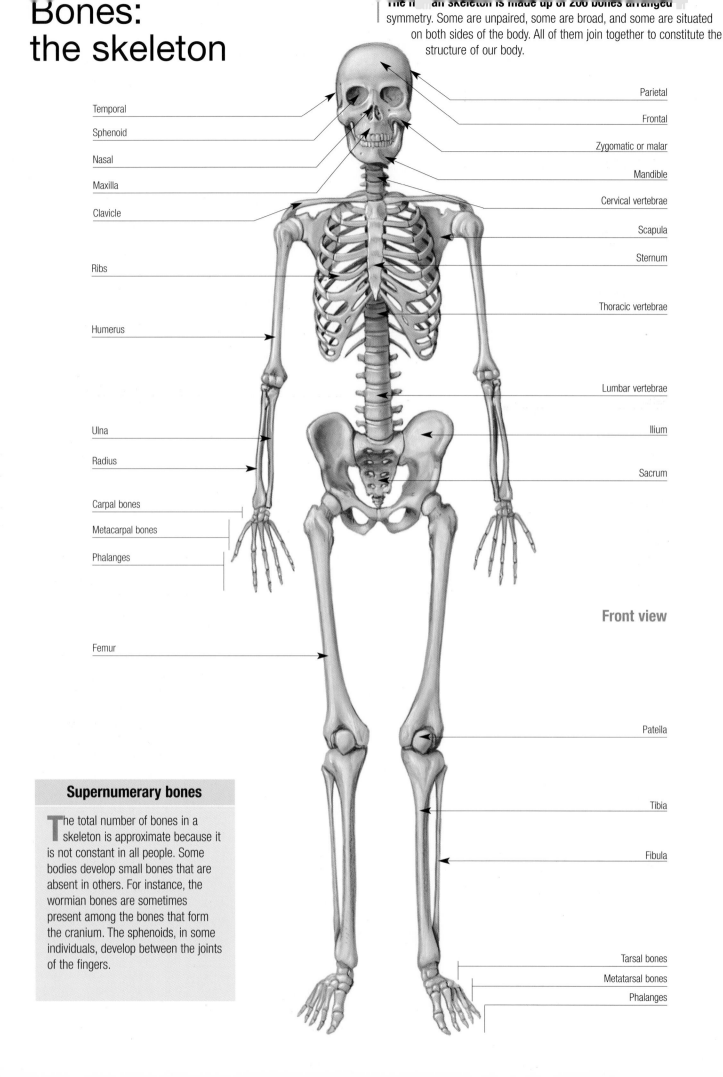

Temporal

Sphenoid

Nasal

Maxilla

Clavicle

Ribs

Humerus

Ulna

Radius

Carpal bones

Metacarpal bones

Phalanges

Femur

Parietal

Frontal

Zygomatic or malar

Mandible

Cervical vertebrae

Scapula

Sternum

Thoracic vertebrae

Lumbar vertebrae

Ilium

Sacrum

**Front view**

Patella

Tibia

Fibula

Tarsal bones

Metatarsal bones

Phalanges

## Supernumerary bones

The total number of bones in a skeleton is approximate because it is not constant in all people. Some bodies develop small bones that are absent in others. For instance, the wormian bones are sometimes present among the bones that form the cranium. The sphenoids, in some individuals, develop between the joints of the fingers.

## Dorsal view

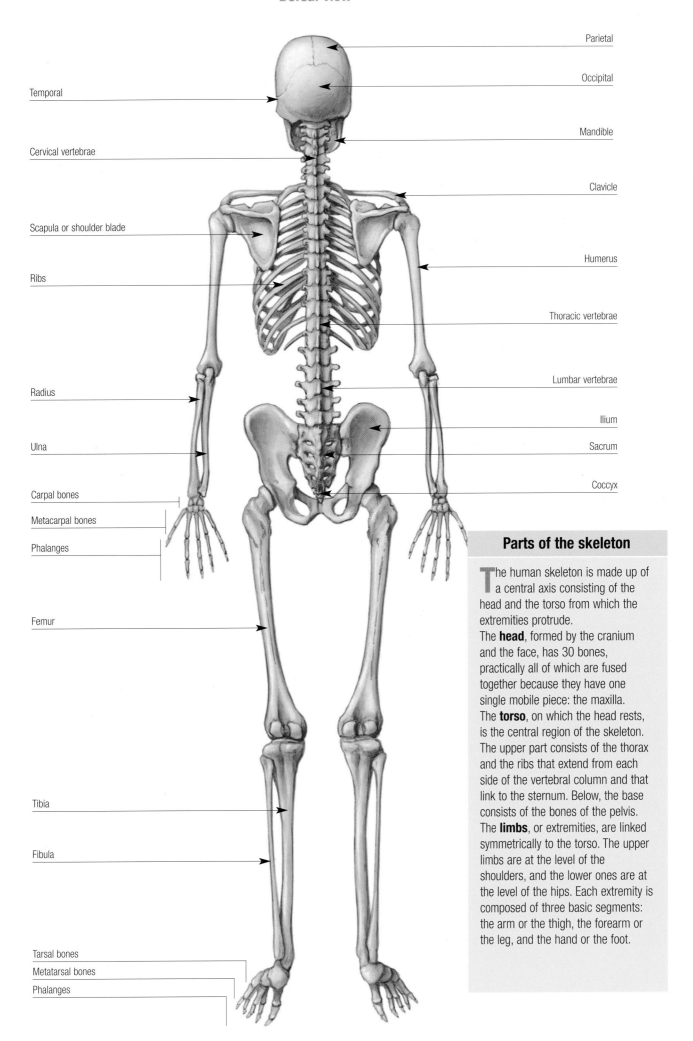

Temporal

Cervical vertebrae

Scapula or shoulder blade

Ribs

Radius

Ulna

Carpal bones

Metacarpal bones

Phalanges

Femur

Tibia

Fibula

Tarsal bones

Metatarsal bones

Phalanges

Parietal

Occipital

Mandible

Clavicle

Humerus

Thoracic vertebrae

Lumbar vertebrae

Ilium

Sacrum

Coccyx

### Parts of the skeleton

The human skeleton is made up of a central axis consisting of the head and the torso from which the extremities protrude.

The **head**, formed by the cranium and the face, has 30 bones, practically all of which are fused together because they have one single mobile piece: the maxilla.

The **torso**, on which the head rests, is the central region of the skeleton. The upper part consists of the thorax and the ribs that extend from each side of the vertebral column and that link to the sternum. Below, the base consists of the bones of the pelvis.

The **limbs**, or extremities, are linked symmetrically to the torso. The upper limbs are at the level of the shoulders, and the lower ones are at the level of the hips. Each extremity is composed of three basic segments: the arm or the thigh, the forearm or the leg, and the hand or the foot.

# The bones
# of the head

is head is composed of 30 bones that form two different parts.
The cranium, which refers to the superior and posterior part, is
composed of eight bones that are fused together and shelters the brain.
The face, which accounts for the inferior and anterior part, is composed
of various bones covered with muscles that form part of the respiratory
and digestive systems.

## Cranium

### Front View

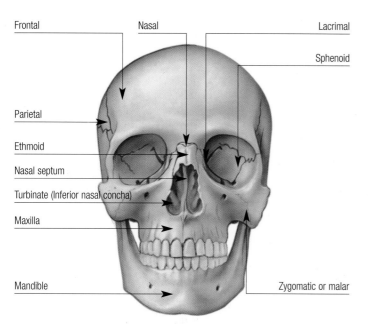

Frontal
Nasal
Lacrimal
Sphenoid
Parietal
Ethmoid
Nasal septum
Turbinate (Inferior nasal concha)
Maxilla
Mandible
Zygomatic or malar

### Side view

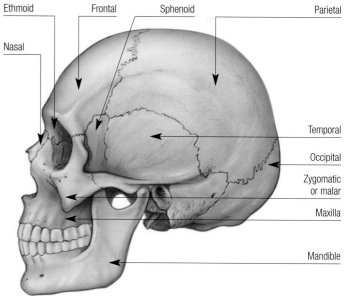

Ethmoid
Frontal
Sphenoid
Parietal
Nasal
Temporal
Occipital
Zygomatic
or malar
Maxilla
Mandible

### Superior view

Frontal
Parietal
Occipital

### Posterior view

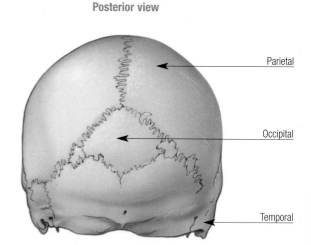

Parietal
Occipital
Temporal

## Bones of the cranium

- Frontal (unpaired bone)
- Parietal (paired bone)
- Temporal (paired bone)
- Occipital (unpaired bone)
- Sphenoid (unpaired bone)
- Ethnoid (unpaired bone)

## Facial bones

- Maxilla (unpaired bone)
- Mandible (unpaired bone)
- Zygomatic or malar (paired bone)
- Vomer (unpaired bone)
- Nasal or nose bone (unpaired bone)

- Turbinates (Inferior nasal conchae) (paired bone)
- Lacrimal (paired bone)
- Palatine (paired bone)
- Hyoid (paired bone)
- Anvil (paired bone)
- Hammer (paired bone)
- Stirrup (paired bone)

## The smallest bones

The head contains the smallest bones of the body: the anvil, the hammer, and the stirrup, which are the tiny bones situated in the middle ear.

## Bones of the head

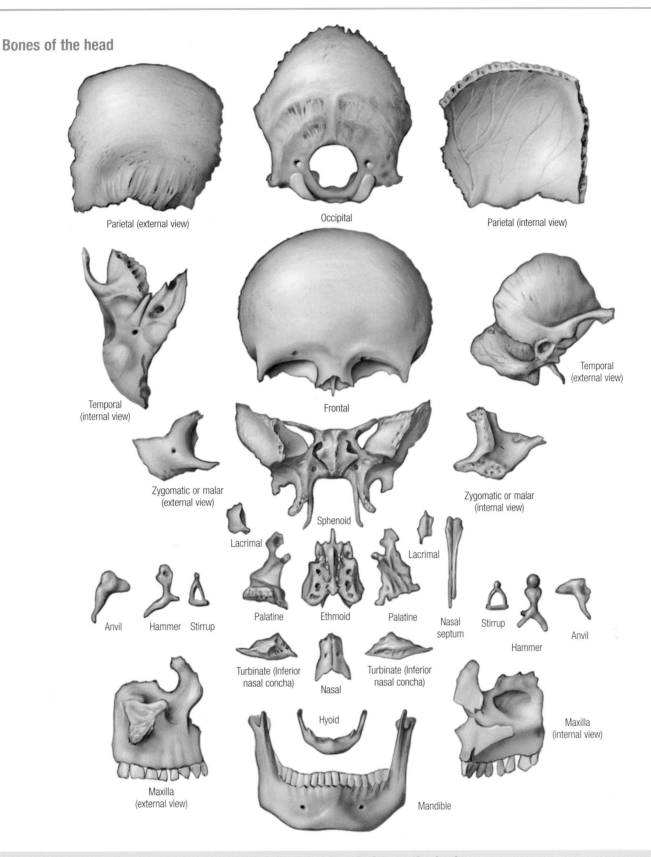

Parietal (external view)

Occipital

Parietal (internal view)

Temporal (internal view)

Frontal

Temporal (external view)

Zygomatic or malar (external view)

Zygomatic or malar (internal view)

Sphenoid

Lacrimal

Lacrimal

Anvil  Hammer  Stirrup

Palatine  Ethmoid  Palatine

Nasal septum

Stirrup

Hammer

Anvil

Turbinate (Inferior nasal concha)

Nasal

Turbinate (Inferior nasal concha)

Maxilla (external view)

Hyoid

Mandible

Maxilla (internal view)

## Growth of the head in relation to the body

At birth, the head is considerably larger compared with the size of the body. It accounts for almost one-third of the body. At the end of pregnancy, the head is the largest part of the body. It is the first to emerge at the moment of delivery. This ratio soon changes. At one year, it accounts for only one-quarter and by adulthood it makes up one-eighth of the body.

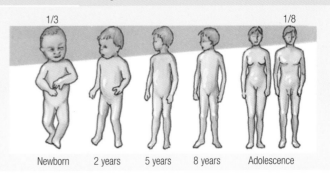

1/3                                                    1/8

Newborn  2 years  5 years  8 years  Adolescence

# The bones of the trunk

human body where the head and limbs are located, consists of the spinal column, the ribs, the sternum, the shoulder blades or scapulas, the clavicles, the ilium or pelvis, and the coccyx.

## Vertebral column

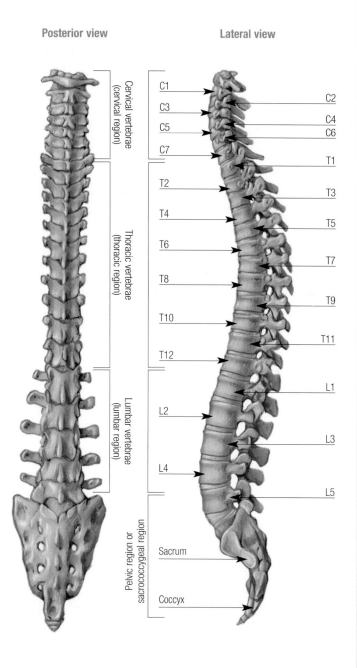

Posterior view

Lateral view

Cervical vertebrae (cervical region)

Thoracic vertebrae (thoracic region)

Lumbar vertebrae (lumbar region)

Pelvic region or sacrococcygeal region

C1
C3
C5
C7
T2
T4
T6
T8
T10
T12
L2
L4

C2
C4
C6
T1
T3
T5
T7
T9
T11
L1
L3
L5

Sacrum

Coccyx

*The vertebral column extends along the entire center line of the back from the base of the cranium to the pelvis. It consists of a series of bones that are superimposed upon each other, the vertebrae. There are a total of 34 vertebrae, although only the 24 upper ones are independent bones. The last ones are fused together and form the sacrum and coccyx bones. The spinal column is divided from top to bottom into four sectors: the cervical region, which consists of the area of the neck; the thoracic or dorsal region, which consists of the area of the thorax; the lumbar region, which consists of the lower portion of the back; and the sacrococcygeal region, which is made up of the sacrum and the coccyx.*

## Vertebrae

These are bones with a rather complex shape. All differ, but all have common features. Each vertebra consists of a compact mass, the vertebral body, and some prolongations called spinous processes. The spinous processes point to the back, and the transverse processes and articular processes point to the sides. The processes are connected to the vertebral body by some osseous pedicles so that they form the spinal canal through which the spinal cord extends along the entire vertebral column.

### Structure of a vertebra: dorsal vertebra

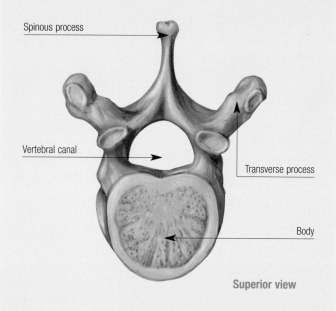

Spinous process

Vertebral canal

Transverse process

Body

**Superior view**

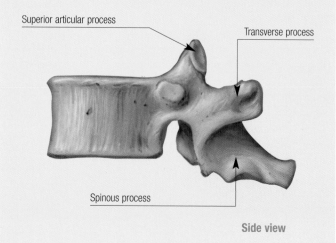

Superior articular process

Transverse process

Spinous process

**Side view**

## Atlas and axis

The first two cervical vertebrae are so special that they have their own names. The first cervical vertebra is called the atlas in honor of the mythological Titan who carried the world on his shoulders. It is joined to the base of the cranium with the occipital bone in such a way that it can support the head. The second cervical vertebra is called the axis. Its principal peculiarity is in the upper part. It has a vertical protrusion that is used as an axis around which the atlas can turn. This makes it possible to turn the head to one side and the other.

### Atlas (superior view)

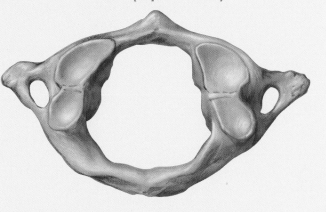

### Axis (posterior superior view)

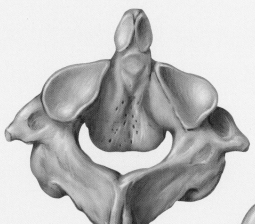

### Bones of the thorax

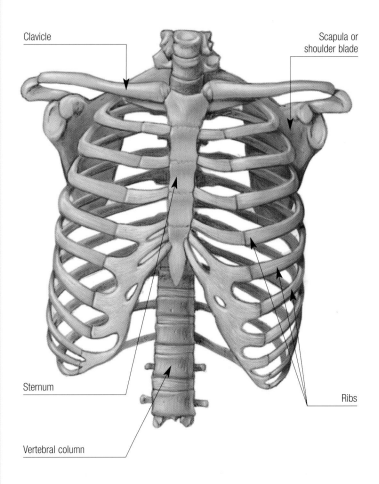

Clavicle

Scapula or shoulder blade

Sternum

Ribs

Vertebral column

### Bones of the pelvis

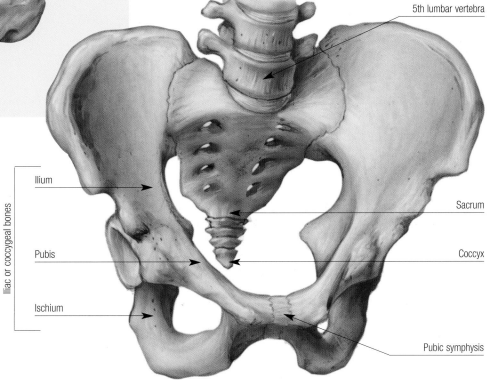

5th lumbar vertebra

Ilium

Iliac or coccygeal bones

Pubis

Ischium

Sacrum

Coccyx

Pubic symphysis

*The iliac or coccygeal bones are located in the lower part of the trunk. These are two symmetrical bones that are joined together in front and with the sacrum in the rear so that a space called the pelvic cavity is formed. Actually, each iliac bone consists of three bones that merge during the growth phase: the ilium, the ischium, and the pubis, linked to the front at a point known as the pubic symphysis.*

# The bones of the upper limb

The skeleton of the upper limb consists of a group of bones that are connected in a very particular way. The only bone of the upper arm is called the humerus. The two bones of the forearm are the ulna and the radius. The various bones of the hand are the carpals, the metacarpals, and the phalanges.

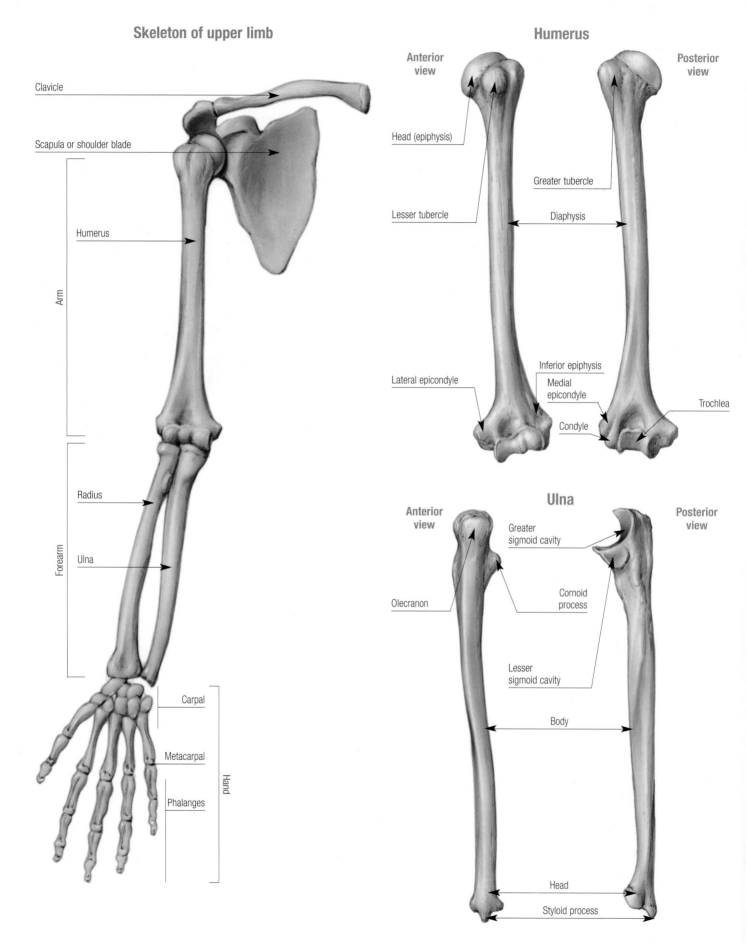

## Skeleton of upper limb

Clavicle

Scapula or shoulder blade

Arm

Humerus

Radius

Forearm

Ulna

Carpal

Metacarpal

Hand

Phalanges

## Humerus

Anterior view

Posterior view

Head (epiphysis)

Greater tubercle

Lesser tubercle

Diaphysis

Lateral epicondyle

Inferior epiphysis

Medial epicondyle

Condyle

Trochlea

## Ulna

Anterior view

Posterior view

Greater sigmoid cavity

Olecranon

Cornoid process

Lesser sigmoid cavity

Body

Head

Styloid process

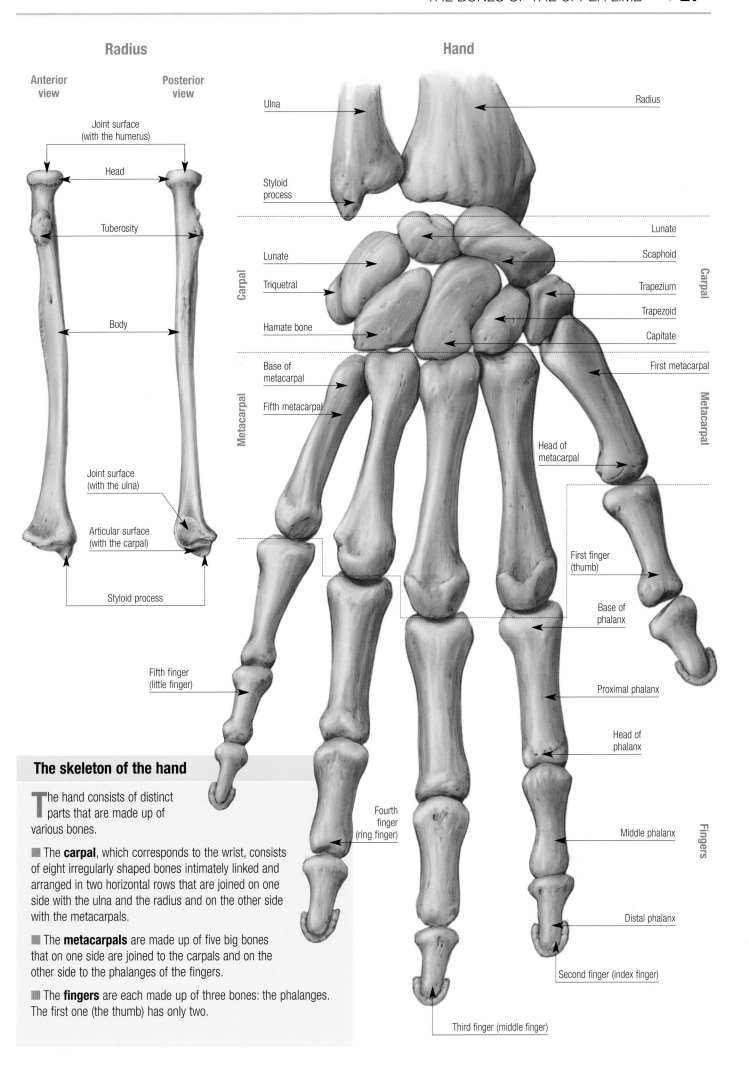

## Radius

**Anterior view**

**Posterior view**

Joint surface (with the humerus)

Head

Tuberosity

Body

Joint surface (with the ulna)

Articular surface (with the carpal)

Styloid process

## Hand

Ulna

Styloid process

Carpal

Lunate

Triquetral

Hamate bone

Metacarpal

Base of metacarpal

Fifth metacarpal

Radius

Lunate

Scaphoid

Trapezium

Trapezoid

Capitate

Carpal

First metacarpal

Metacarpal

Head of metacarpal

First finger (thumb)

Base of phalanx

Fifth finger (little finger)

Fourth finger (ring finger)

Proximal phalanx

Head of phalanx

Middle phalanx

Fingers

Distal phalanx

Second finger (index finger)

Third finger (middle finger)

### The skeleton of the hand

The hand consists of distinct parts that are made up of various bones.

■ The **carpal**, which corresponds to the wrist, consists of eight irregularly shaped bones intimately linked and arranged in two horizontal rows that are joined on one side with the ulna and the radius and on the other side with the metacarpals.

■ The **metacarpals** are made up of five big bones that on one side are joined to the carpals and on the other side to the phalanges of the fingers.

■ The **fingers** are each made up of three bones: the phalanges. The first one (the thumb) has only two.

# The bones of the lower limb

The skeleton of the lower limb consists of components that interconnect in a very particular way. The only bone in the thigh is the femur. The two bones of the lower leg are the tibia and the fibula. The various bones of the foot are the tarsals, the metatarsals, and the toes.

## Skeleton of lower limb

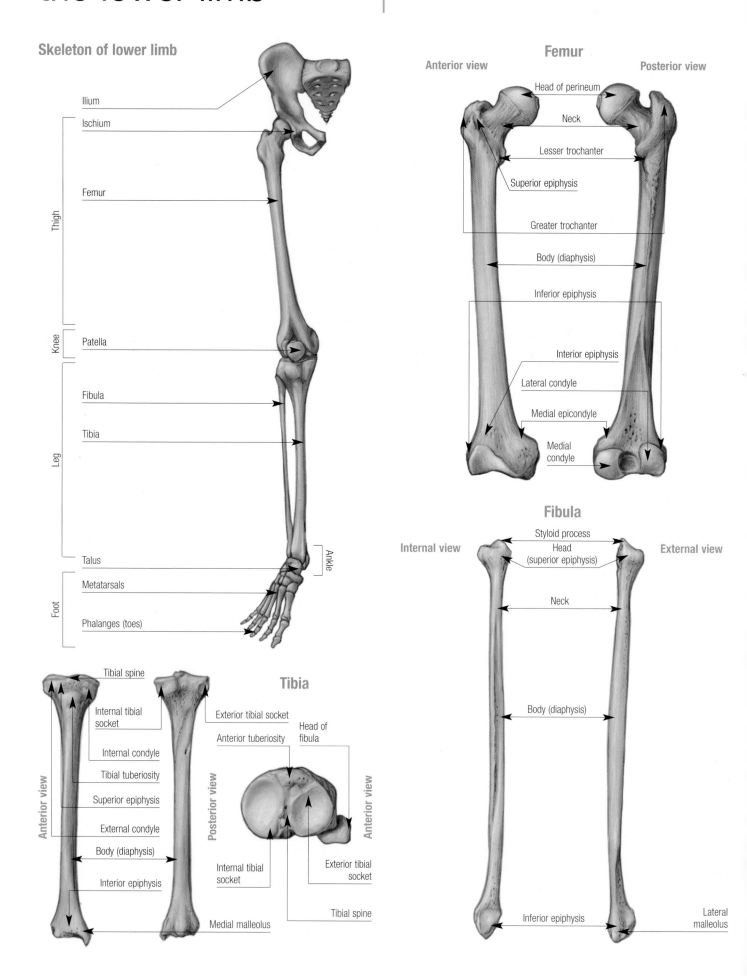

Ilium
Ischium
Femur
Thigh
Patella
Knee
Fibula
Tibia
Leg
Talus
Ankle
Metatarsals
Foot
Phalanges (toes)

## Femur

Anterior view
Posterior view

Head of perineum
Neck
Lesser trochanter
Superior epiphysis
Greater trochanter
Body (diaphysis)
Inferior epiphysis
Interior epiphysis
Lateral condyle
Medial epicondyle
Medial condyle

## Fibula

Internal view
External view

Styloid process
Head (superior epiphysis)
Neck
Body (diaphysis)
Inferior epiphysis
Lateral malleolus

## Tibia

Anterior view
Posterior view
Anterior view

Tibial spine
Internal tibial socket
Internal condyle
Tibial tuberiosity
Superior epiphysis
External condyle
Body (diaphysis)
Interior epiphysis
Medial malleolus

Exterior tibial socket
Anterior tuberiosity
Head of fibula
Internal tibial socket
Exterior tibial socket
Tibial spine

## The patella, a single bone

The patella is a special bone that is part of the knee joint. The major role of the patella is to protect the knee from trauma. It moves only during flexion and extension.

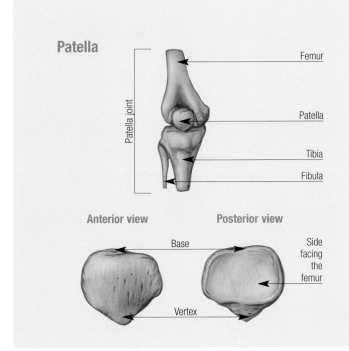

Patella

Patella joint

Femur

Patella

Tibia

Fibula

**Anterior view**

Base

Vertex

**Posterior view**

Side facing the femur

## Foot skeleton

The foot is composed of separate parts each consisting of various bones.

■ The **tarsus**, the posterior part of the foot that includes the heel, is composed of seven closely linked bones arranged in two rows, one of which is joined to the tibia and the fibula.

■ The **metatarsus**, which consists of the instep and the sole of the foot, consists of five large bones.

■ The **toes** each have three bones, the phalanges, except for the first toe (the big toe), which has only two.

## Bones of the foot

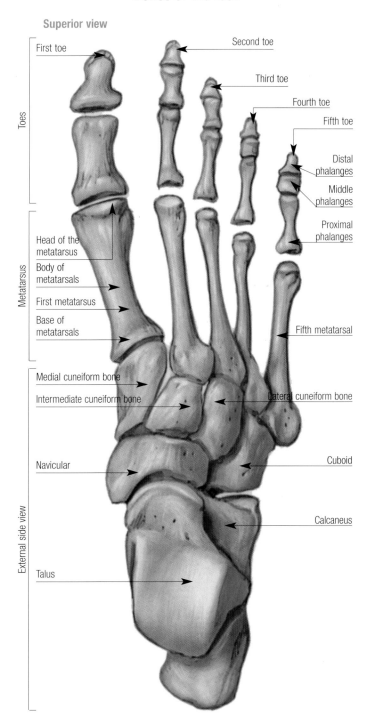

**Superior view**

Toes

First toe

Second toe

Third toe

Fourth toe

Fifth toe

Distal phalanges

Middle phalanges

Proximal phalanges

Metatarsus

Head of the metatarsus

Body of metatarsals

First metatarsus

Base of metatarsals

Fifth metatarsal

Medial cuneiform bone

Intermediate cuneiform bone

Lateral cuneiform bone

External side view

Navicular

Cuboid

Calcaneus

Talus

**External side view**

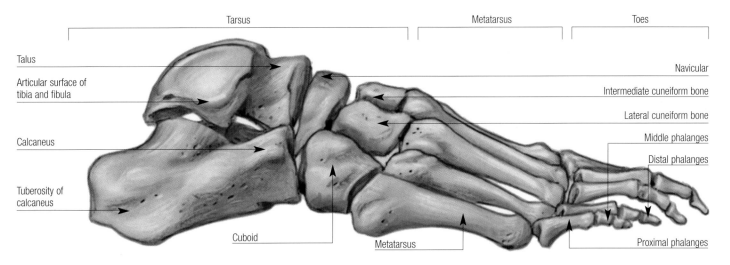

Tarsus

Metatarsus

Toes

Talus

Articular surface of tibia and fibula

Calcaneus

Tuberosity of calcaneus

Cuboid

Metatarsus

Navicular

Intermediate cuneiform bone

Lateral cuneiform bone

Middle phalanges

Distal phalanges

Proximal phalanges

# The muscles: general aspects

**The skeletal muscles consist of fleshy masses that contract,** thereby changing length. They are inserted in the bones and other structures by fibrous bands called tendons. The contractions of the muscles are responsible for the movements of our body.

## Structure of a Skeletal Muscle

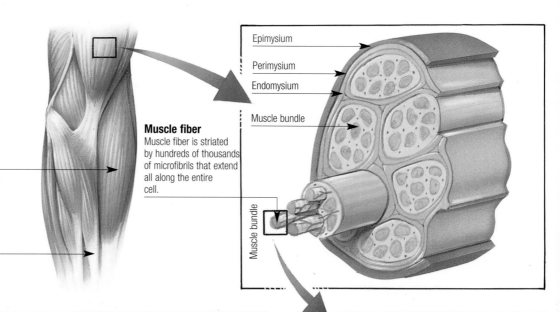

**Muscle**
Muscle is made up of a group of elongated cells that are called muscle fibers. They are arranged in bundles surrounded by strands of connective tissue.

Tendon

**Muscle fiber**
Muscle fiber is striated by hundreds of thousands of microfibrils that extend all along the entire cell.

Epimysium

Perimysium

Endomysium

Muscle bundle

Muscle bundle

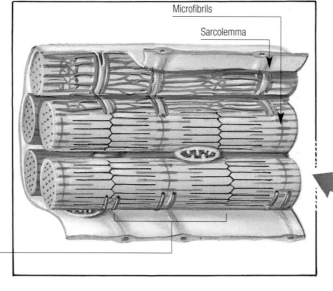

**Sarcomere**
It contains intertwined actin and myosin protein filaments. When they receive the appropriate nerve stimuli, they slide on top of each other, resulting in muscle contraction.

Microfibrils

Sarcolemma

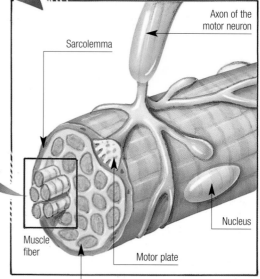

Sarcolemma

Axon of the motor neuron

Nucleus

Muscle fiber

Motor plate

**Microfibril**
When observed under the electron microscope, the microfibril has a series of regular microfilaments that form bands of different colors. These in turn, form the functional units of the muscle called the sarcomeres.

### Microscopic structure of a microfibril

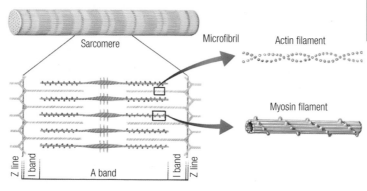

*Each sarcomere is bounded by dark striations called Z lines. Inside it has central striations with a darker tone, called A bands, and two brighter transverse striations, called I bands. These bands indicate the presence of actin filaments, which are thicker, and myosin, which are thinner and intertwined with each other. When the muscle fiber is relaxed, the contact surface between both types of protein filaments is at a minimum. When the fiber receives a nerve stimulus, the thin filaments slide over the thick filaments, causing the distance*

*between the Z lines, which makes up the boundary of the sarcomere, to become narrow. This, in turn, reduces the length of the microfibrils, and the stimulated muscle fiber becomes*

*shorter. When the stimulus stops, the actin–myosin filaments resume their prior position and the muscle is relaxed.*

Sarcomere

Microfibril

Actin filament

Myosin filament

Z line
I band
A band
I band
Z line

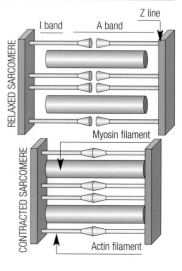

I band | A band | Z line

RELAXED SARCOMERE

Myosin filament

CONTRACTED SARCOMERE

Actin filament

## Muscular contraction

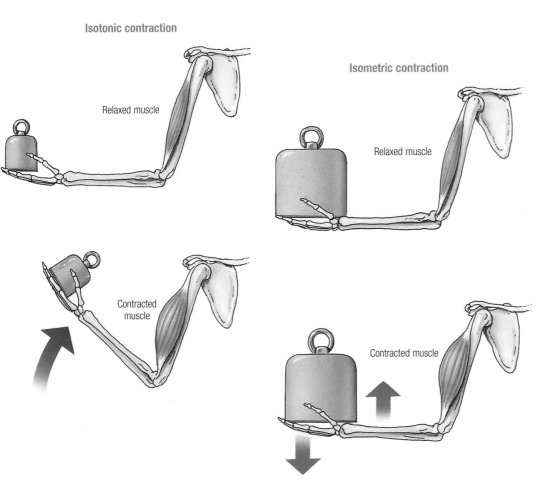

### Isotonic contraction

Relaxed muscle

Contracted muscle

### Isometric contraction

Relaxed muscle

Contracted muscle

*Muscles become tense in response to a specific nerve stimulus. If the internal tension is greater than the resistance of the force of gravity or a specific body part, the muscle will contract and the distance between its points of attachment will become shorter. Movement therefore takes place. This type of muscle contraction, or **isotonic contraction,** is responsible for all body movements. If the internal tension generated in the stimulated muscle is less than the tension created by the force of gravity or a certain body segment, then the muscle "swells up." The distances between its points of attachment do not become narrower, and there is no movement. This phenomenon, which is known as **isometric contraction,** is responsible for the muscle tone that maintains the balance of the body as a whole and of the body segments in various positions.*

## Agonist and antagonist muscles

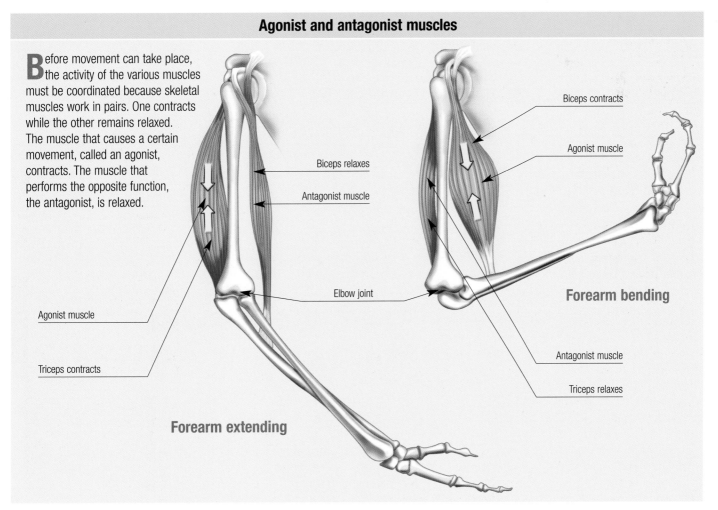

**B**efore movement can take place, the activity of the various muscles must be coordinated because skeletal muscles work in pairs. One contracts while the other remains relaxed. The muscle that causes a certain movement, called an agonist, contracts. The muscle that performs the opposite function, the antagonist, is relaxed.

Biceps relaxes

Antagonist muscle

Elbow joint

Agonist muscle

Triceps contracts

**Forearm extending**

Biceps contracts

Agonist muscle

**Forearm bending**

Antagonist muscle

Triceps relaxes

# The muscles of the body

The human body has about 600 different skeletal muscles. Some are longer, and others are shorter. They vary in size and ability. Muscles are responsible for the movements of the body as a whole and of each of the separate parts of the body.

## Anterior view of the muscles of the human body

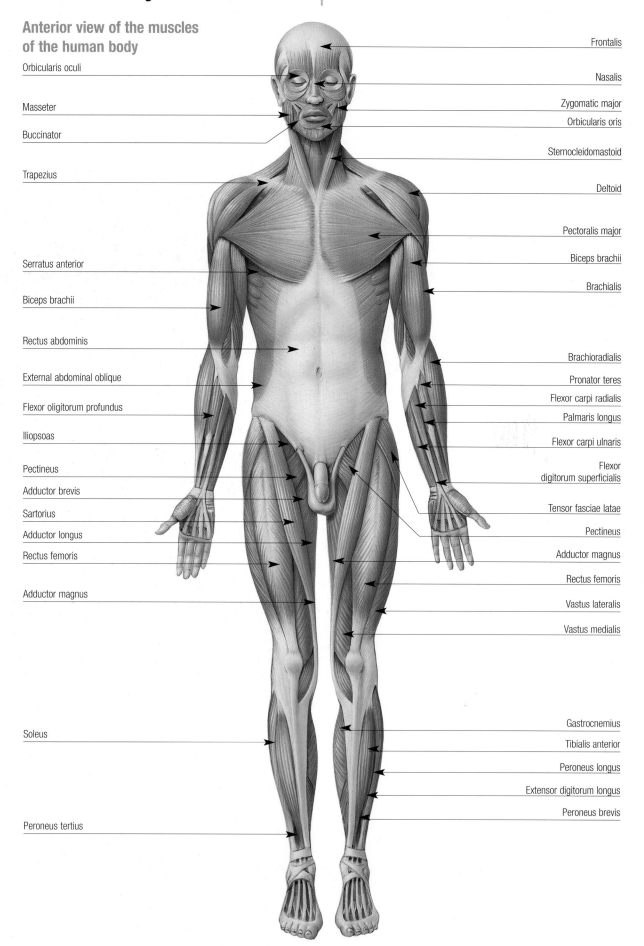

Orbicularis oculi

Masseter

Buccinator

Trapezius

Serratus anterior

Biceps brachii

Rectus abdominis

External abdominal oblique

Flexor oligitorum profundus

Iliopsoas

Pectineus

Adductor brevis

Sartorius

Adductor longus

Rectus femoris

Adductor magnus

Soleus

Peroneus tertius

Frontalis

Nasalis

Zygomatic major

Orbicularis oris

Sternocleidomastoid

Deltoid

Pectoralis major

Biceps brachii

Brachialis

Brachioradialis

Pronator teres

Flexor carpi radialis

Palmaris longus

Flexor carpi ulnaris

Flexor digitorum superficialis

Tensor fasciae latae

Pectineus

Adductor magnus

Rectus femoris

Vastus lateralis

Vastus medialis

Gastrocnemius

Tibialis anterior

Peroneus longus

Extensor digitorum longus

Peroneus brevis

## Posterior view of the muscles of the human body

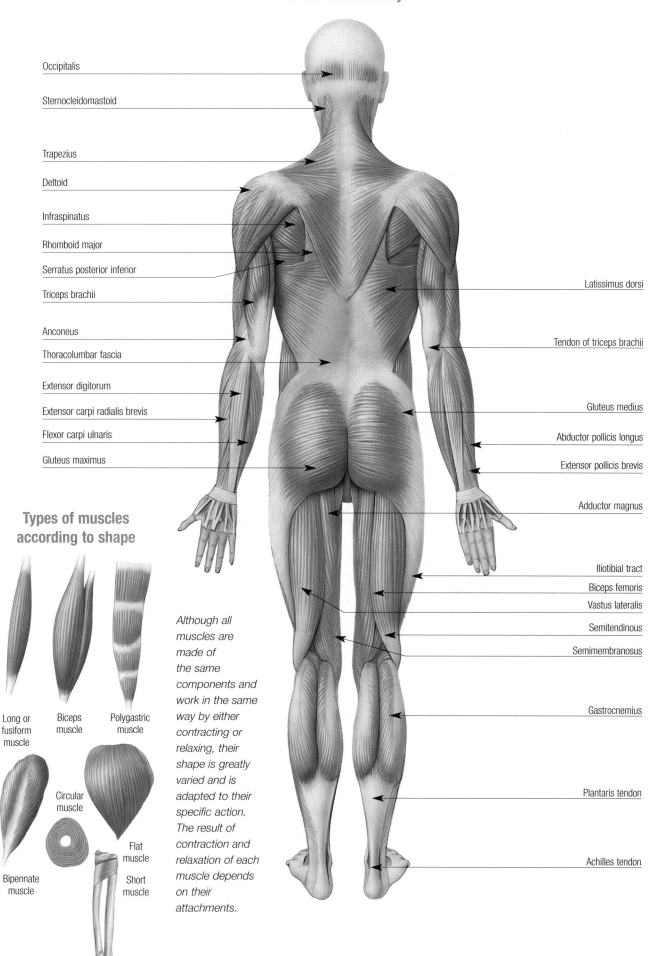

Occipitalis

Sternocleidomastoid

Trapezius

Deltoid

Infraspinatus

Rhomboid major

Serratus posterior inferior

Triceps brachii

Anconeus

Thoracolumbar fascia

Extensor digitorum

Extensor carpi radialis brevis

Flexor carpi ulnaris

Gluteus maximus

Latissimus dorsi

Tendon of triceps brachii

Gluteus medius

Abductor pollicis longus

Extensor pollicis brevis

Adductor magnus

Iliotibial tract

Biceps femoris

Vastus lateralis

Semitendinous

Semimembranosus

Gastrocnemius

Plantaris tendon

Achilles tendon

## Types of muscles according to shape

Long or fusiform muscle

Biceps muscle

Polygastric muscle

Circular muscle

Flat muscle

Bipennate muscle

Short muscle

*Although all muscles are made of the same components and work in the same way by either contracting or relaxing, their shape is greatly varied and is adapted to their specific action. The result of contraction and relaxation of each muscle depends on their attachments.*

# Joints: types

**The joints are the contact points between the bones of the** skeleton. Soft elements, such as cartilage and synovial fluid, cushion the space between bones. There are different types of joints. Some are fixed, but most of them are more or less mobile, each having its own special function.

## Types of joints

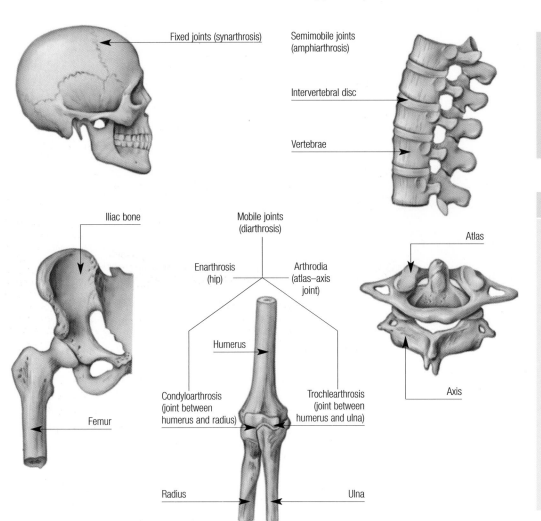

Fixed joints (synarthrosis)

Semimobile joints (amphiarthrosis)

Intervertebral disc

Vertebrae

Iliac bone

Mobile joints (diarthrosis)

Enarthrosis (hip)

Arthrodia (atlas–axis joint)

Atlas

Humerus

Condyloarthrosis (joint between humerus and radius)

Trochlearthrosis (joint between humerus and ulna)

Axis

Femur

Radius

Ulna

The human body has about 200 joints that facilitate the movements of the various parts of the skeleton and enable people to move from one place to the other.

## Menisci

Some mobile joints have elastic, fibrocartilaginous structures that promote the coupling of some bone segments that do not fit well together. These are the menisci. The biggest ones, and also those most frequently used, are the menisci of the knee. Other joints also have this type of structure, such as the temporomandibular joint, the sternoclavicular joint, and the acromioclavicular joint.

## Types of joints

We distinguish three basic types of joint depending on their degree of mobility: fixed, semimobile, and mobile.

**Fixed joints (synarthrosis).** Devoid of any movement, fixed joints consist of the solid union of two or more bone segments. Their primary task is to constitute a protective layer for the soft tissues, as in the case of the joints of the cranial bones, which protect the brain.
**Semimobile joints (amphiarthrosis).** With semimobile joints, the bone surfaces are not directly tied together. Instead, they are separated by a fibrocartilaginous structure that permits only slight movements. The vertebral joints are separated from each other only by an intervertebral disc. Although each joint has little mobility, the spinal column can bend forward or incline toward the sides.
**Mobile joints (diarthrosis).** Mobile joints allow a vast range of movements. These include the joints of the limbs, such as the shoulder, the hip, the elbow, and the knee. We distinguish various

types according to their shape and how the bone segments fit together.
**Enarthrosis:** Enarthroses joints are made up of spherical bone segments that fit within a cavity. They can move in all directions. A chief example is the hip joint, which relates to the linkage of the head of the femur to the iliac bone.
**Condyloarthroses:** Condyloarthroses joints are commonly called ball-and-socket joints. One example is the joint of the radius with the humeral condyle.
**Trochlearthrosis:** Trochlearthrosis joints link a bone segment in the shape of a pulley that has a depression in the center and another bone segment that consists of a crest that fits into the pulley channel, such as the joint of the ulna with the humerus.
**Arthrodia:** Arthrodial joint faces are smooth and flat, which is why they can slide only among each other, such as in the case of the first two cervical vertebrae.

## Types of mobile joints

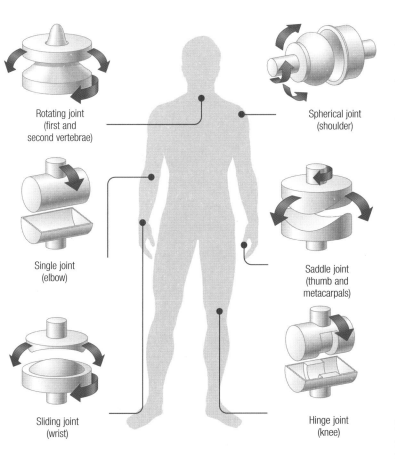

Rotating joint
(first and
second vertebrae)

Single joint
(elbow)

Sliding joint
(wrist)

Spherical joint
(shoulder)

Saddle joint
(thumb and
metacarpals)

Hinge joint
(knee)

*We distinguish various types of mobile joints, depending on their shape and the way the bone segments that they link fit together; each facilitates specific movements.*

## Components of a mobile joint (shoulder)

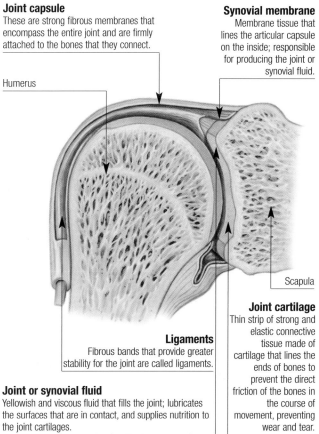

**Joint capsule**
These are strong fibrous membranes that encompass the entire joint and are firmly attached to the bones that they connect.

Humerus

**Synovial membrane**
Membrane tissue that lines the articular capsule on the inside; responsible for producing the joint or synovial fluid.

Scapula

**Ligaments**
Fibrous bands that provide greater stability for the joint are called ligaments.

**Joint cartilage**
Thin strip of strong and elastic connective tissue made of cartilage that lines the ends of bones to prevent the direct friction of the bones in the course of movement, preventing wear and tear.

**Joint or synovial fluid**
Yellowish and viscous fluid that fills the joint; lubricates the surfaces that are in contact, and supplies nutrition to the joint cartilages.

*In addition to the related bone segments in each particular case, the mobile joints contain other tissues and elements that are indispensable in order to achieve correct function and provide the necessary stability for the joint.*

## Movements of the shoulder

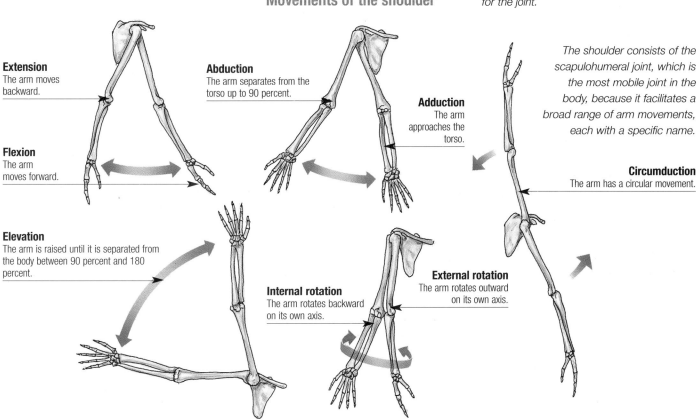

**Extension**
The arm moves backward.

**Flexion**
The arm moves forward.

**Elevation**
The arm is raised until it is separated from the body between 90 percent and 180 percent.

**Abduction**
The arm separates from the torso up to 90 percent.

**Adduction**
The arm approaches the torso.

**Internal rotation**
The arm rotates backward on its own axis.

**External rotation**
The arm rotates outward on its own axis.

**Circumduction**
The arm has a circular movement.

*The shoulder consists of the scapulohumeral joint, which is the most mobile joint in the body, because it facilitates a broad range of arm movements, each with a specific name.*

# Joints: problems

**The joints are structures that are more complex than they seem** and, at the same time, are fragile. They must support the friction that results from movements. In many cases, they must also carry the load of the body weight. Therefore, joint problems are frequent, and, in many cases, they develop over a long period of time.

## Evolution of arthritis

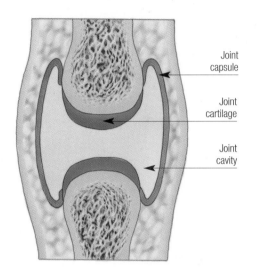
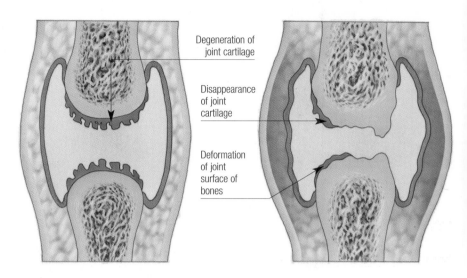

Joint capsule

Joint cartilage

Joint cavity

Degeneration of joint cartilage

Disappearance of joint cartilage

Deformation of joint surface of bones

## Arthritis

Arthritis is a chronic disorder due to the progressive degeneration of the joint cartilage that, under normal conditions, prevents the direct friction of the bony ends of the mobile joints and protects them against wear and tear. As this cartilage loses its elasticity and becomes thin, the bone ends are in direct contact and, therefore, undergo progressive deterioration that impairs the function of the stricken joint. The disorder develops slowly and progressively so that joint lesions appear silently and without showing any symptoms over a long period of time. When the joint cartilage is greatly deteriorated, almost destroyed, its absence impairs the integrity of the rest of the joint structures. The consequences include joint pain and stiffness as well as restricted movements.

## Locations of arthritis

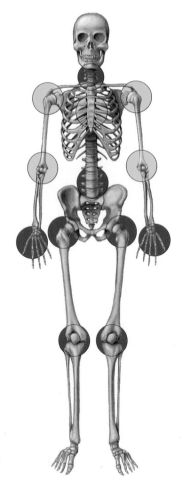

### Arthritis and obesity

**B**ody weight is not a direct cause of arthritis. However, it considerably promotes arthritic development because it causes an overload for the affected joints.

*Arthritis can occur in practically all mobile joints of the body, but those that are stricken with greater frequency are the knee, the hand, the hip, and the vertebral column, especially in the cervical and lumbar regions.*

● Most-frequent locations

○ Less-frequent locations

*Arthritis of the cervical spine can cause extreme pain in the region of the neck that requires the temporary use of an orthopaedic collar.*

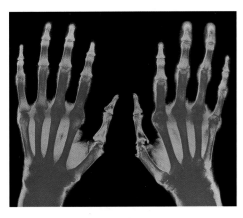

*X-ray of the hands of a person suffering from rheumatoid arthritis*

## Rheumatoid arthritis

Rheumatoid arthritis is a chronic disease that symmetrically affects the various joints, especially those of the limbs. Its specific origin is not known, although we do know that it is an autoimmune disorder. The immune system reacts in an abnormal fashion and produces antibodies against the tissues in the body itself, in this case against the components of the joints, which become inflamed and lose their ability to function. The disease develops in a manner characterized by alternating periods of acuteness and remission. In the long run, in the most serious cases, it leads to real disability.

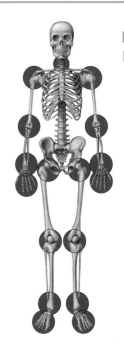

**Locations of Rheumatoid Arthritis**

## Home adaptations for people with rheumatoid arthritis

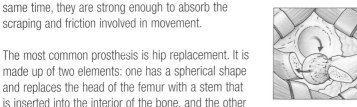

### In the bath

Safety bar attached to ceiling

Bathtub safety rails

Adjustable bathtub seat

Button-operated water tank

Safety bars

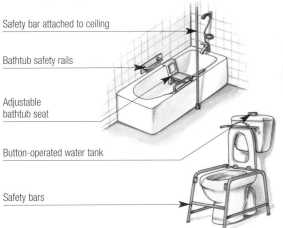

### In the kitchen

Handles on big and heavy doors

Double-handled pots

Broom and big dust pans

Furniture lower than usual

Seats with back rest and foot support

Handles for cups and glasses

Nonslip surfaces

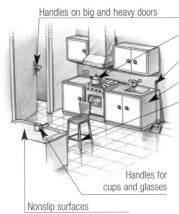

*The illustrations show some of the adjustments that must be made in the kitchen and the bathroom in the home of those afflicted with advanced rheumatoid arthritis in order to ease the inconveniences connected with progressive loss of joint function.*

The replacement of bone segments of a joint by artificial prostheses is a very useful step in dealing with pain and incapacitating functional failures from illnesses such as arthritis or advanced rheumatoid arthritis. Today, many types of prostheses are made of metal alloys and plastic materials that are perfectly tolerated by the body because they do not generate any rejection reaction by the immune system. At the same time, they are strong enough to absorb the scraping and friction involved in movement.

The most common prosthesis is hip replacement. It is made up of two elements: one has a spherical shape and replaces the head of the femur with a stem that is inserted into the interior of the bone, and the other one has a concave form that is implanted in the acetabulum of the iliac bone, attached with a special cement. Because both parts are perfectly coupled together, the joint can then perform all of its normal movements.

### Hip prosthesis

Iliac bone

Ball

Stem

Femur

Cavity

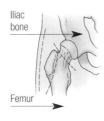

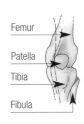

### Knee prosthesis

Femur

Patella

Tibia

Fibula

Femoral component

Tibial component

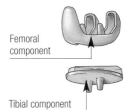

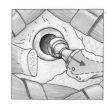
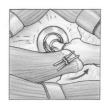
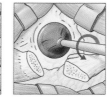
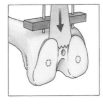
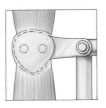

# Deformations of the feet

**The feet, which support the entire body weight, can suffer** deformations that can have multiple repercussions on the skeleton. Even though they may not be dangerous, they certainly cause discomfort. Some are typical of childhood, such as flat feet, and others are more associated with adults, such as hallux valgus, which causes bunions.

## Normal arches of the sole

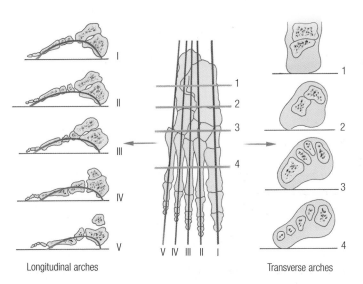

Longitudinal arches

Transverse arches

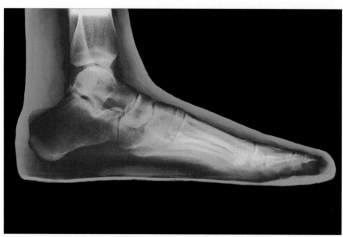

*An obvious example of flat feet*

## Flat feet

Flat feet contribute to the absence or the lowering of the normal sole curvature, which can lead to local discomfort and, even more importantly, can have repercussions on other parts of the skeleton. The foot is a complex structure that supports body weight while a person stands up and walks. Under normal conditions, the sole is arched in the shape of a half moon that is open toward the internal side of the foot. The body weight is then distributed among various supporting points. In the case of flat feet, the arch is lower in height than normal or is absent, and turns the heel outward.

## Repercussions

Initially, deformation causes discomfort only if a person remains on his or her feet much of the time or after prolonged walking. During that stage, the problem can be solved by simple treatment intended to train and strengthen the musculature of the feet. If nothing is done, the defect is accentuated over the years. The foot bones eventually adjust to the deformation, which becomes permanent. During that phase, it is customary for pain to develop in the feet when walking a short time, spreading to the legs. On the other hand, the lowering of the arch can cause the knees to become deviated. If lowering of the arch becomes more accentuated by one foot rather than the other, the length of both limbs will differ, causing a limp and a compensatory deviation of the vertebral column.

## Exercises recommended to correct flat feet

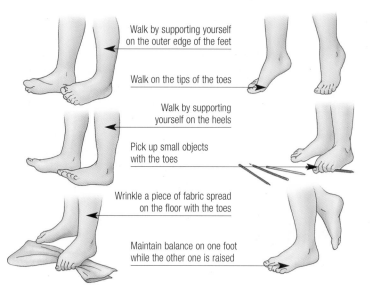

Walk by supporting yourself on the outer edge of the feet

Walk on the tips of the toes

Walk by supporting yourself on the heels

Pick up small objects with the toes

Wrinkle a piece of fabric spread on the floor with the toes

Maintain balance on one foot while the other one is raised

## Sole imprints

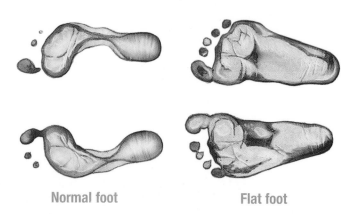

Normal foot

Flat foot

*Sole imprints are very useful in detecting flat feet because they reveal abnormal support of the sole in an area that is normally not in contact with the surface.*

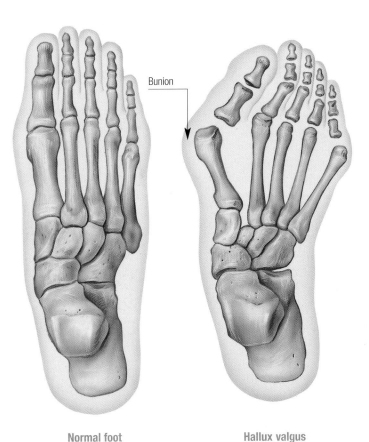

**Normal foot**          **Hallux valgus**

### *Hallux valgus* (bunion)

*Hallux valgus* is a deformation of the big toe where the tip of the toe is deviated toward the other toes. The appearance of a typical prominence at its base is called a bunion. A bunion causes painful discomfort. Under normal conditions, the big toe, whose Latin name is *hallux*, is perfectly lined up with the first metatarsal bone.

*While the metatarsaphalangeal joint retains its mobility, one can still slow the development of a bunion and diminish discomfort by means of some small rods that will keep the structures of the foot in the proper position. After a real deviation of the bones has already taken place, the only effective treatment is surgery.*

## Development

The deviation of the big toe usually develops slowly and progressively so that, in most cases, it becomes evident only toward the fourth decade of life. In the first phase, the principal symptom is alternating pain in the metatarsus. This is usually an uncomfortable but not too intensively painful condition that appears especially when one stands too long on the feet or after a lengthy walk. This pain gets worse when one wears tight shoes and high heels, but diminishes and disappears when at rest. Later on, when the deformation is already more in place, the pain is usually intense and constant. It does not readily stop even when at rest.

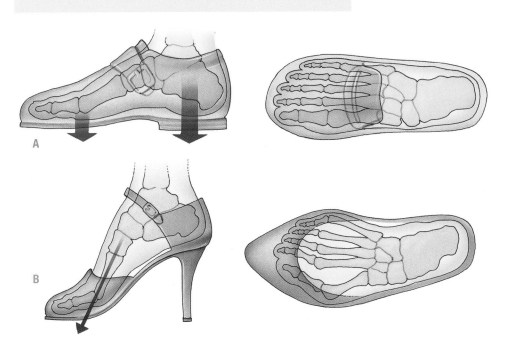

*Deformation is usually a consequence of improper shoe wear. Short shoes with a pointed tip force the toe toward the axis of the foot, against the other toes. Shoes with high heels also lead to anatomical deformation. With appropriate shoe wear (A), body weight is distributed correctly over the entire foot. In shoes with high heels (B), body weight is supported by the front half of the foot and, if the shoe has a pointed tip, pressure is exerted against the toe, leading to the formation of the bunion.*

# Cervical and lumbar pain

**Pain in the neck and lumbar region is a principal cause of work** absenteeism. The vertebral column is subjected to stresses, incorrect posture, and abrupt movements that frequently cause muscular and skeletal changes in these regions.

## Pain in the neck

Causes of cervical pain differ widely because the structures of the area are varied. Mechanical problems related to disorders cause irritation of the nerve endings located in the joint, ligaments, and osseous and muscular structures of the cervical spine. Contraction of the neck muscles caused by stress creates discomfort.

Acute cervical pain is characterized by the abrupt appearance of pain in the neck that sometimes radiates toward the nape, the shoulders, or the anterior part of the thorax. Its intensity varies, but on many occasions it is very pronounced and gets worse as a result of any kind of movement. It can be accompanied by a reflective contraction of the musculature of the region that limits the mobility of the neck. Most of the time, it lasts several days or,

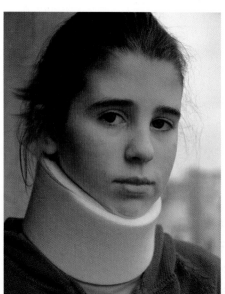

in some cases, several weeks, before going into remission until it disappears.

Chronic cervical pain can also begin quite abruptly as the result of an acute problem that does not go into remission, although it most frequently occurs in a progressive manner. It is usually a persistent, intermittent, or fluctuating pain of variable intensity.

*The use of an orthopaedic collar is an excellent way to immobilize the neck, and often this is sufficient to ease cervical pain.*

## Torticollis

**A** painful contraction of the muscles of the neck that prevents the normal movement of the entire region and that locks the head in a fixed position or prevents it from rotating to one side is called torticollis. This disorder usually comes on quite suddenly. It is often triggered by abrupt traumatic movements. Sleeping in a bad position can bring it on. Generally, there is a spasmodic contraction of one of the sternocleidomastoid muscles. The head remains locked in a fixed position to one side, and any movement toward the other side turns out to be painful or impractical. Treatment is based on keeping the area at rest. Using an orthopaedic collar, along with muscle relaxants as well as massage therapy or physiotherapy, helps to alleviate the condition.

### Muscles of the neck region

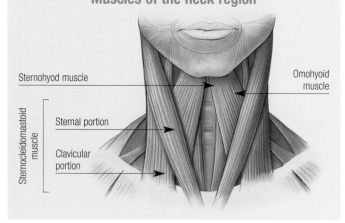

Sternohyod muscle

Omohyoid muscle

Sternocleidomastoid muscle

Sternal portion

Clavicular portion

## Cervicobrachial neuralgia

**I** rritation or compression of nerve roots that emerge from the spinal cord in the cervical region of the spinal column can give rise to intense pain in the neck, the upper part of the back, and the upper limbs. The exact location of pain depends on the root, in other words, the region it innervates. The most common cause is a herniated disc as the result of a trauma, or it can spring from a chronic degenerative process. The pain usually occurs abruptly and can be accompanied by other neurological symptoms in the stricken area, such as a sensation of pins and needles, a sensation of cold or hot, a prickly sensation, or cramps. In general, discomfort ends spontaneously within several weeks, but the most important thing is to keep the area at rest. Only in severe cases is it necessary to resort to surgery to bring about decompression and repair the responsible lesions.

### Location of pain in cervicobrachial neuralgia

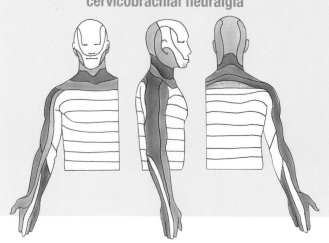

## Herniated disc

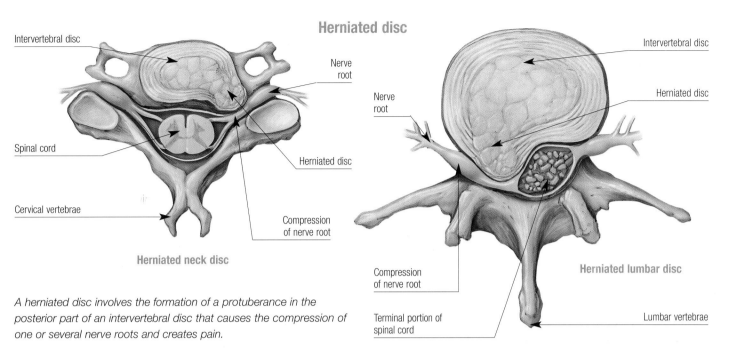

Intervertebral disc

Nerve root

Spinal cord

Herniated disc

Cervical vertebrae

Compression of nerve root

**Herniated neck disc**

Intervertebral disc

Herniated disc

Nerve root

Compression of nerve root

Terminal portion of spinal cord

**Herniated lumbar disc**

Lumbar vertebrae

*A herniated disc involves the formation of a protuberance in the posterior part of an intervertebral disc that causes the compression of one or several nerve roots and creates pain.*

### Lumbago and sciatica

The term "lumbago" indicates acute pain in the lumbar part of the back. Although there are many and varied causes, the disorder is usually due to a musculoskeletal change. It is usually caused by an irritation or compression of the nerves that originate in the lumbar region of the spinal cord. The nerves absorb major pressures while standing and are subject to herniation of an intervertebral lumbar disc, causing pressure against a spinal root. The pain is usually very intense and is located in the lumbar region and sometimes radiates toward the buttocks and the pelvis. Compression may take place against any of the lumbar and sacral roots that form the sciatic nerve.

## Sciatic nerve

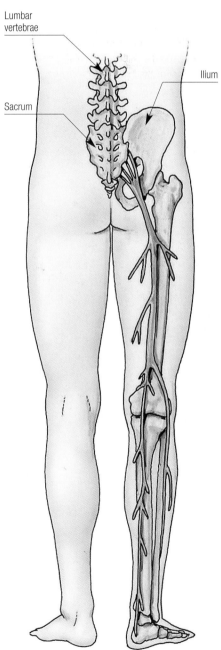

Lumbar vertebrae

Ilium

Sacrum

### Locations of Pain and Sciatica

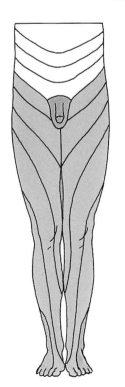
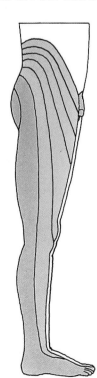

# Prevention of back pain

**Prevention of back pain is based on avoiding exaggerated** movements, extremely heavy weights, and sudden movements that affect the vertebral column, the delicate structure that is the axis of our skeleton. Strains of the vertebral column are the main cause of back pain.

## Movements and activities that can cause lumbar pain

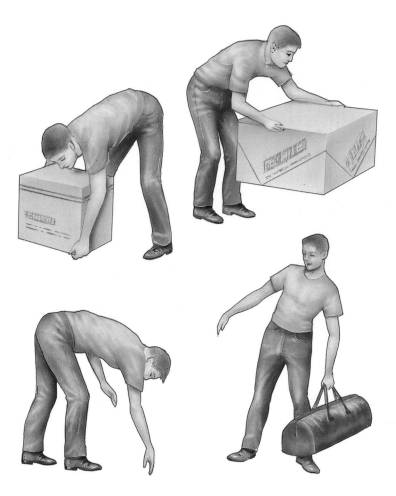

*Poor posture when reading or writing for a long time is one of the most common causes of back pain. Having correct posture when performing these activities, therefore, is the main means of prevention.*

*Attacks of back pain are caused by activities that entail an effort or an overload in the lumbar region of the spinal column. Pain will be triggered when an abrupt or improper activity is performed. When one inclines forward without bending the knees, lifts a big and heavy object without help, or transports a very heavy object with only one hand, major back pain may be a result.*

### It is a good idea:

- To maintain an adequate body weight.
- To sleep on a mattress that is not too soft and a wooden box spring. Slide a slat between the box spring and the mattress to correct a soft mattress.
- To sleep on your sides to avoid back pain.
- To move to the end of the bed and extend your legs before sitting up.
- To sit down and put your foot on an elevated support to put on your shoes.
- To walk in shoes with 2-inch (5 cm) or 2½ inch (6 cm) heels.
- To stand on a stool to reach an object that is in a high position.
- To turn the entire body to pick up any object situated behind you.
- To bend the legs to pick up a heavy weight.
- To carry any heavy load as close as possible to the body.
- To distribute the weight of any object between both arms.
- To push any heavy object by putting your back against it.
- To sit down with your back straight and resting on a backrest. Make sure your knees are even with your hips and your feet are well planted on the ground.
- To sit in seats with armrests and good backrests.
- To take rests when you do extensive work.

### It is not a good idea:

- To be overweight.
- To sleep on a soft bed.
- To sleep with your head on a low pillow.
- To sit up abruptly while your legs are extended.
- To bend down to put on your shoes.
- To wear shoes with heels that are either too high or too low.
- To stretch to reach an object in a high position.
- To turn only your torso to reach an object behind you.
- To bend your torso but not your knees to lift a heavy object.
- To carry a heavy object away from your body.
- To carry a heavy object with only one arm.
- To push a heavy object from the front.
- To sit down with your back bending forward.
- To sit down in chairs that are too soft.
- To stay in the same position for a long time.

## Advice to prevent back pain

| Bad | Good | Bad | Good |
| --- | --- | --- | --- |

## Exercises recommended to prevent cervical pain

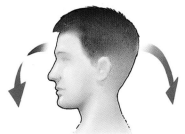

Flexion and extension of neck

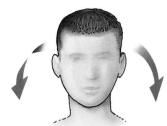

Sideways inclination of head

Turning the head to the right and left

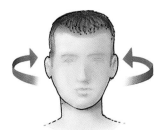

Rotation of arms, making a circle

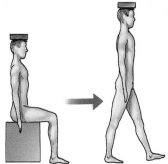

Standing from a chair and walking with an object on the head.

### "Dangerous" activities

**N**umerous activities involve positions that cause exaggerated stress on one or more parts of the spinal column and can cause back pain. Examples are architects and tailors who sit in one position for a long time, leaning forward. They usually have pain between the shoulder blades. Truck drivers, because of constant movements and vibrations, become accustomed to pain in the area of the kidneys. Gardeners who bend their backs instead of bending the legs to squat commonly complain of pain in the lower part of the back. When performing an activity involving improper positions of the back, proper precautions should be taken to minimize the risks.

*Exercises intended to guarantee the mobility of the neck and to strengthen the muscles in this area are extremely useful in protecting the cervical spine. Proper exercises must be performed regularly. Effort must be gradual and progressive because any abruptness or excessive effort will turn out to be counterproductive.*

# Sprains and luxations

**Sprains and luxation dislocations are very common. They** frequently occur as a result of accidents during sports or work activities. Proper first aid can prevent the worsening of these injuries, lead to prompt recovery, and prevent their recurrence.

## Mechanisms that can cause ankle sprains

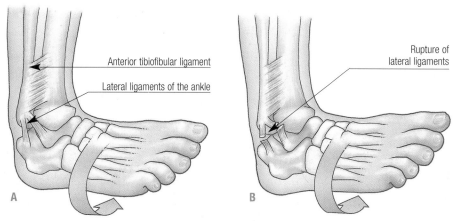

Anterior tibiofibular ligament

Lateral ligaments of the ankle

Rupture of lateral ligaments

A

B

*Turning the foot inward can cause the ankle to be twisted due to distension (A) or tear (B) of the lateral ligaments of that joint.*

## Sprains

A sprain involves a split, a rupture, or an avulsion (tearing away) of the joint ligaments as a result of exaggerated movement. The cause can be a very abrupt movement or a trauma, a twist, or a simple stumble. The joint is forced to perform a movement that it is not capable of performing. Sprains can occur in any joint, but the most frequent location is the ankle, which supports a good part of the body's weight. Sprains also occur in the knees and the fingers, most commonly in relation to sports accidents.

The earliest symptom is pain that occurs immediately and is usually intense. It prevents any kind of movement of the stricken joint, and the ankle is involved even when the foot is on the ground. The stricken joint is inflamed and swells up, while the skin that covers it appears red and hot. It is not unusual to see bruises in this area, indicating a vascular lesion and the subsequent hemorrhages that accompany lesions of the ligaments.

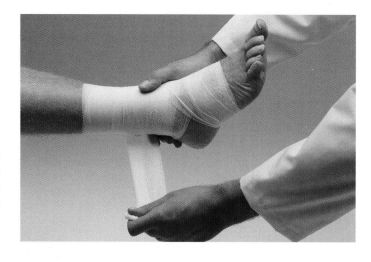

*Immobilization of the stricken joint is the principal therapeutic measure for a sprain.*

## ✚ First aid

First, avoid any movement of the stricken joint. Place it in an elevated position. To counteract the immediate inflammation and the internal hemorrhage that accompany the injury, apply cold to the site. Cold should be applied at most for half an hour, although this procedure can be repeated when the area becomes warm again.

It is best to seek medical assistance to rule out a fracture. Always be sure the joint is kept at rest. Avoid any forced movements during transfer to a health facility. If one cannot get prompt medical assistance, apply a compression bandage, such as an ace bandage, to ensure complete immobilization.

**1.** Keep the joint immobile and in an elevated position.

**2.** Apply cold compresses or an ice pack to the joint for a maximum of a half hour.

**3.** Apply a slightly compressive bandage, and seek medical help.

## Most frequent dislocations

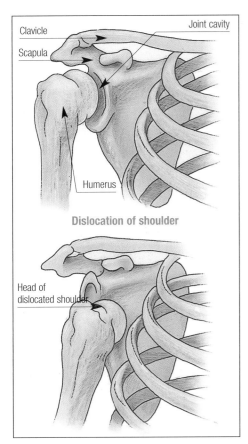

Clavicle
Scapula
Joint cavity
Humerus

**Dislocation of shoulder**

Head of dislocated shoulder

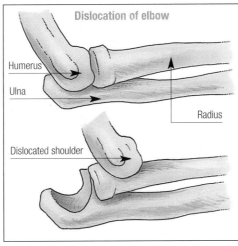

**Dislocation of elbow**

Humerus
Ulna
Radius

Dislocated shoulder

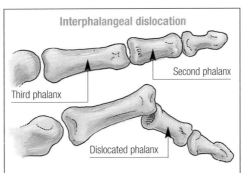

**Interphalangeal dislocation**

Third phalanx
Second phalanx
Dislocated phalanx

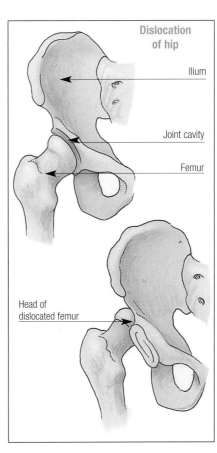

**Dislocation of hip**

Ilium
Joint cavity
Femur

Head of dislocated femur

Practically any joint can suffer a dislocation. This problem is most frequent in certain locations, particularly in the limbs. The most common dislocation is that of the shoulder. Next in terms of frequency are dislocations of the elbow and the fingers. These occur most frequently in the course of sports activities, such as baseball, basketball, and football. Dislocations of the hip occur less frequently. They are usually the result of a fall.

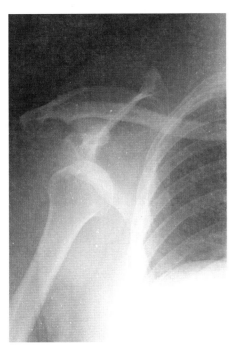

X-ray of a dislocation of the right shoulder showing the displacement of the humerus head.

*Generally, dislocations are manifested as a deformity of the stricken joint associated with swelling and pain in the area.*

### Luxation

A luxation or dislocation involves displacement of the bone segments that form a joint, usually accompanied by sprains of ligaments and the joint capsule. The most common cause is a violent trauma that involves failure of the restraining elements of the joint and displacement of a bone. This can be caused by either a direct or an indirect trauma to a bone or joint.

The earliest symptom is pain that appears immediately after the accident and that impedes or completely prevents moving the stricken joint. Depending on the type of dislocation and the degree of displacement of the bone segment and its position, it may become impossible to perform some or all of the movements of the injured joint. The joint may also become deformed.

### Appearance of a dislocated joint

Dislocation of first metacarpal

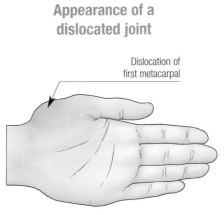

Normal shoulder
Dislocated shoulder

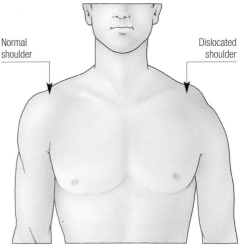

# Fractures

**A bone fracture involves the rupture or interruption of the** continuity of the bone and may be partial or complete. It most commonly occurs as a result of a fall or a trauma that causes a violent impact. A fracture can be caused by any kind of trauma or even an abrupt movement.

## Type of fractures

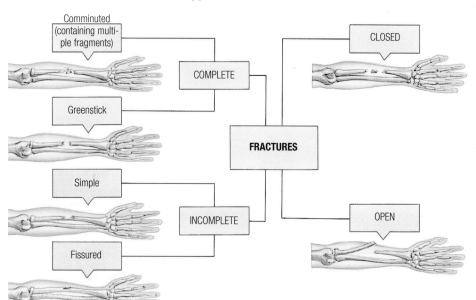

Comminuted (containing multiple fragments)

COMPLETE

Greenstick

FRACTURES

Simple

INCOMPLETE

Fissured

CLOSED

OPEN

*The various types of fractures are classified depending upon the degree of rupture of the bone (complete, when the bone is broken into two parts or more, or incomplete, when there is a partial loss of bone continuity) and according to whether or not the bone ends are visible outside the body (closed, if the skin that covers the area of the broken bone remains intact, or open, when the surface tissues are torn and the bone fragments remain in direct contact with the outside).*

## Symptoms

Generally, the fracture is inflamed and an internal hemorrhage can give rise to the appearance of a more or less extensive hematoma. If the bone fragments are displaced, an obvious deformation whose size varies depending on the position of these fragments can usually be detected. Sometimes there is a loss of total function, and it is completely impossible to perform any kind of movement. In other cases, the skeletal segment involved can display abnormal mobility. Other manifestations depend on the possible complications.

### Possible fracture complications

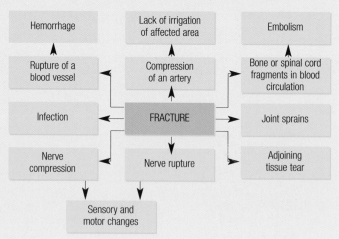

| Hemorrhage | Lack of irrigation of affected area | Embolism |
| Rupture of a blood vessel | Compression of an artery | Bone or spinal cord fragments in blood circulation |
| Infection | FRACTURE | Joint sprains |
| Nerve compression | Nerve rupture | Adjoining tissue tear |

Sensory and motor changes

## Action to be taken in case of forearm fracture before going to a hospital

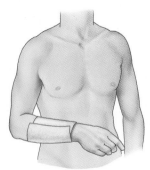

Cover injured area with a cloth

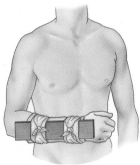

Immobilize the forearm with splints

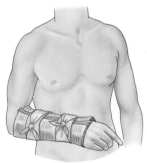

Add other rigid elements

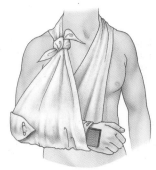

Put on a sling to support the forearm

## Improvised immobilization of fractures of the lower extremity

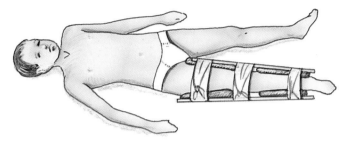

Immobilization of a fracture of the leg with splints and pieces of cloth

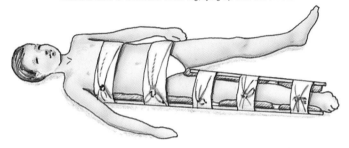

Immobilization of a fracture of the leg, tying up both extremities

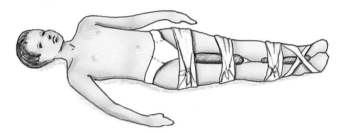

Immobilization of a fracture of the femur with splints and pieces of cloth

## Action to be taken in case of open fracture before going to a hospital

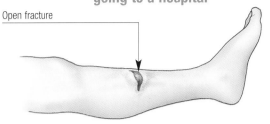

Open fracture

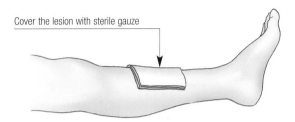

Cover the lesion with sterile gauze

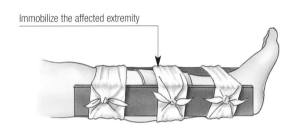

Immobilize the affected extremity

*In dealing with an open fracture, stopping the hemorrhage and preventing infection is as important as immobilization. Before immobilizing the area, therefore, it is necessary to cover the wound with a clean dressing, preferably with sterile gauze.*

## Improvised sling

Improvised sling using a short shirt. The garment is folded on the injured side so that it will surround the forearm, and the free portion is fixed at the level of the chest with safety pins.

Sling improvised with a longer shirt. Unbutton the last two buttons and fold the part on the injured side in such a way that it will surround the forearm, knotting the end with a piece of cloth surrounding the neck.

*To immobilize an upper limb when appropriate material is not available, use the clothing item carried by the accident victim in order to cover the torso. Basically, with the elbow fixed in position, the forearm is supported against the torso and surrounded with the clothing item. Attach the sling in some way.*

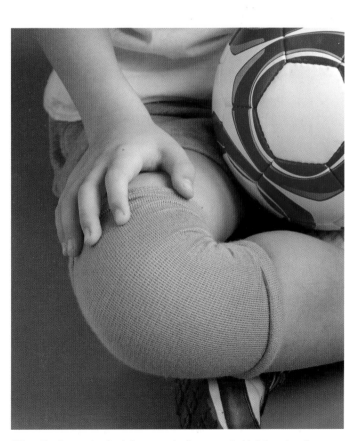

*Often the improvised reinforcement of a wounded joint makes it possible to continue with sports activities, but this usually makes the problem worse or delays recovery.*

# Makeup of blood

**Blood is a red, viscous fluid that runs constantly throughout the** interior of the circulatory system. It is composed of a special fluid, which is called plasma and is in suspension. Blood transports various types of blood cells and an endless number of various substances throughout the body.

## Functions of blood

- Blood carries oxygen and nutrients throughout the entire body.
- It transports waste from metabolism and toxic substances to organs that eliminate or neutralize them.
- It transports the hormones produced by the endocrine glands to the tissues where they perform their actions.
- It participates in thermal adjustments of the body.
- It collaborates with the body's defense system.

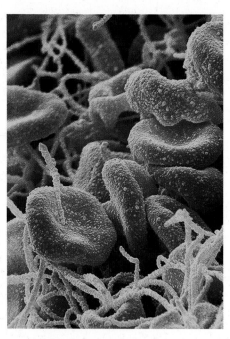

*Red blood cells viewed under the microscope.*

## Basic components of blood

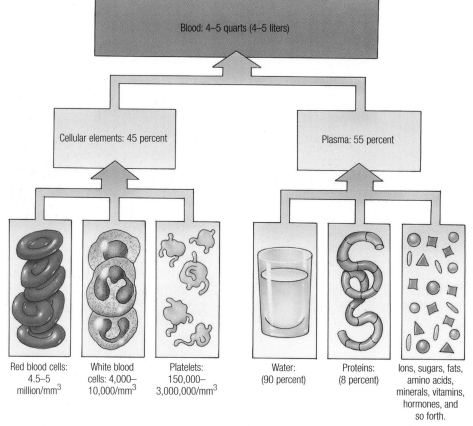

Blood: 4–5 quarts (4–5 liters)

Cellular elements: 45 percent

Plasma: 55 percent

Red blood cells: 4.5–5 million/mm$^3$

White blood cells: 4,000–10,000/mm$^3$

Platelets: 150,000–3,000,000/mm$^3$

Water: (90 percent)

Proteins: (8 percent)

Ions, sugars, fats, amino acids, minerals, vitamins, hormones, and so forth.

**Blood plasma** is a yellowish fluid composed basically of water (90 percent). It carries all of the elements present in the blood throughout the interior of the cardiovascular system. In addition to blood cells, plasma serves as a vehicle for nutrients, mineral substances, metabolic wastes, vitamins, hormones, and multiple products that carry out varied biological functions. Some of these substances travel freely in the plasma. However, many of them are insoluble and form complexes with proteins that they transport in the blood to release in the appropriate part of the body.

**Blood cells** float in the plasma. There are three basic types of cells: red blood cells, whose particular color gives blood its characteristic tone; white blood cells, among which we distinguish diverse varieties; and platelets, the smallest blood cells.

## Red blood cells

Red blood cells, also called erythrocytes, are the most abundant blood cells. They have the shape of a biconcave disc with a diameter of 7.5 microns. In reality, they are incomplete cells because they lack a nucleus, which means that their lifetime is limited to some 120 days. Inside, they contain hemoglobin, a pigment made up of iron and responsible for the color of blood. Hemoglobin enables them to perform their basic function: to transport oxygen from the lungs to the tissues of the body, and carbon dioxide, a residue from metabolism, in the opposite direction back to the lungs, where the blood becomes oxygenated again.

7.5 μm

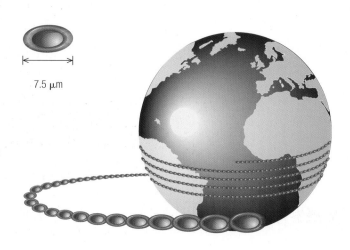

If we were to line up all of the red blood cells of an adult, one after the other, more than 2 billion cells (4.5 million/mm$^3$ x 5 quarts of blood) would run approximately 5.3 times around the equator.

## Types of white blood cells (leukocytes)

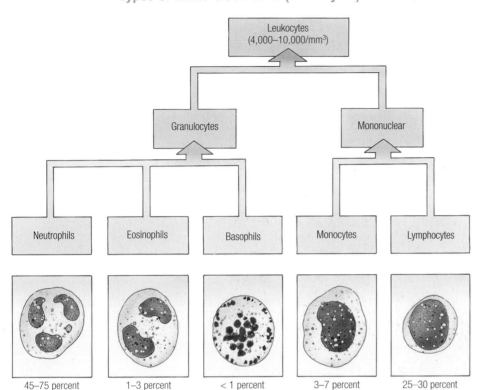

| Leukocytes (4,000–10,000/mm³) | | | | |
|---|---|---|---|---|
| **Granulocytes** | | | **Mononuclear** | |
| Neutrophils | Eosinophils | Basophils | Monocytes | Lymphocytes |
| 45–75 percent | 1–3 percent | < 1 percent | 3–7 percent | 25–30 percent |

## White blood cells

White blood cells, also called leukocytes, are part of the immune system that protects the body against infections. There are various types of white blood cells, all of which have a nucleus and some of which even have a multilobed nucleus that, when viewed under the microscope, appears to be more than just one nucleus. As a result, some leukocytes are called polynuclear and others are called mononuclear.

**Polynuclear leukocytes** are also called **granulocytes** because, under the microscope, one can observe inside them a series of granules containing the substances necessary for their particular functions. They are differentiated into three fundamental types: **neutrophils**, responsible for phagocytosis or "eating" foreign substances, primarily bacteria; **eosinophils**, which are involved in allergic reactions and in parasite infections; and **basophils**, which are involved in allergic reactions. There are two types of **mononuclear leukocytes**: **monocytes**, that are responsible for phagocytosing germs, detritus, and all kinds of foreign elements, and **lymphocytes** that are responsible for producing antibodies (B lymphocytes) and directly attacking the invading microorganism (T lymphocytes).

## Platelets

Platelets, also called thrombocytes, are the smallest blood cell elements. They are incomplete cells and have an average lifespan of some ten days. Their function is focused on stopping hemorrhages because they actively participate in the process of coagulation.

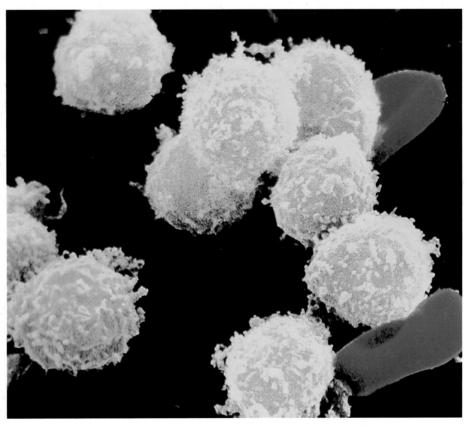

*Stained white blood cells viewed through an electron microscope.*

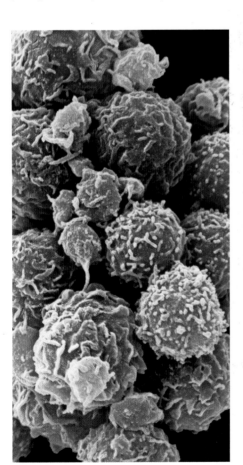

*Platelets viewed through an electron microscope.*

# Formation of blood

**The formation of the cellular elements of blood is a process** called hematopoiesis and takes place in the bone marrow present inside some bones and also, to a lesser extent, in the spleen and the lymph nodes.

## Uninterrupted production

The following are formed every day:

- 100,000 to 250,000 million red blood cells
- 30,000 million white blood cells
- 70,000–150,000 million platelets

## Location of red bone marrow in an adult

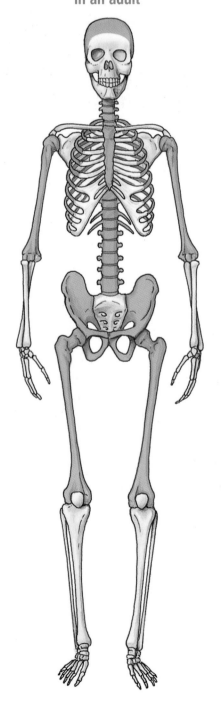

## Bone marrow

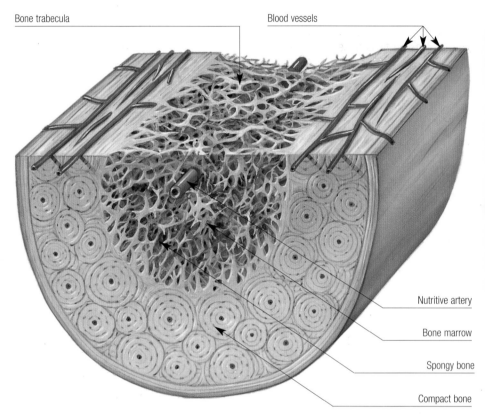

Bone trabecula

Blood vessels

Nutritive artery

Bone marrow

Spongy bone

Compact bone

## Bone marrow

This tissue specializes in the production of the various cell elements of the blood. It is located inside the bones, both in the medullary cavity and also in the trabeculae of the spongy bone tissue situated under the outer layer of the compact bone tissue. Two types of bone marrow are differentiated. Red bone marrow is responsible for producing blood cells, and yellow bone marrow is inactive and very rich in fatty tissue. The proportion and location of both types of bone marrow varies with age. In the newborn, all bones of the skeleton contain red bone marrow. In an adult, there is active red bone marrow only in the ribs, the sternum, the vertebrae, the bones of the cranium, the pelvis, and the distal parts of the long bones.

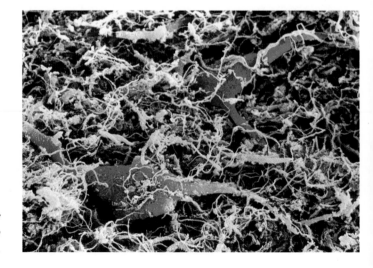

*Bone marrow viewed through an electron microscope.*

The spleen is an organ involved in the production and destruction of blood cells. It has a soft consistency and egg-shaped form. It is situated in the upper left part of the abdomen. Its interior is spongy, and it is grooved by partitions that divide it into several areas. In the central part of the organ, we find the splenic artery through which a large volume of blood, distributed over numerous arterioles, travels until it reaches a multitude of lacunae and then passes to a series of venules that flow together to form the splenic vein.

Around the arterioles, there are some accumulations of lymphoid tissue, called Malpighian cells, that constitute the so-called white pulp, composed of some blood lacunae or venous sinuses, and some trabeculae of reticular tissue called Billroth's cords.

## Spleen

**Ventromedial view of spleen**

Splenic artery

Splenic vein

Hilus

**Side view of spleen**

Fibrous trabeculae

Capsule

Trabecular veins

Trabecular arteries

Hilus

Parenchyma

Splenic vein

Splenic artery

Billroth's strands

Trabecular vein

trabecular vein

Red pulp

Venous sinuses

Trabecular artery

Malpighian cells

White pulp

**Microscopic structure of spleen**

## Hematopoiesis

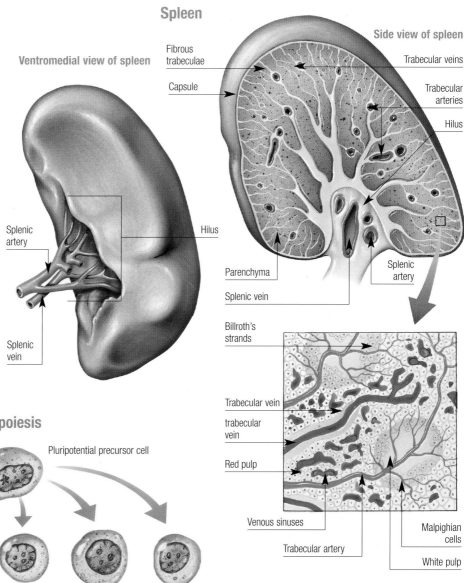

Pluripotential precursor cell

Megakaryoblast

Monopotential precursor cell

Monopotential precursor cell

Monopotential precursor cell

Monopotential precursor cell

Myeloblast

Promonoblast

Lymphoblast

Proerythroblast

Promyelocyte

Erythroblast

Myelocyte

Premonocyte

Prolymphocyte

Megakaryocyte

Reticulocyte

Metamyelocyte

Hook or band

Platelets

Red blood cells

Granulocyte

Monocyte

Lymphocyte

Formation of various blood cells, called hematopoiesis, is uninterrupted. All the elements have a limited lifetime, so they must be constantly replaced. The process basically takes place in the bone marrow where there are some special precursor cells of all the types of blood cells. The pluripotential precursor cells are capable of reproducing by themselves and differentiating themselves to become monopotential mother cells, and are prepared to generate each specific type of blood cell.

From their beginning, the blood elements pass through a process of modulation in various stages that are given different names until finally they are turned into red blood cells, white blood cells, or platelets that then move into circulation.

# Blood groups and transfusions

**Although it always has the same components, blood is classified** based on specific differences among those components. Understanding the differences among blood types is necessary to determine if blood from one individual can be transfused into another without endangering the life of the recipient.

## Frequency of blood groups in Caucasians

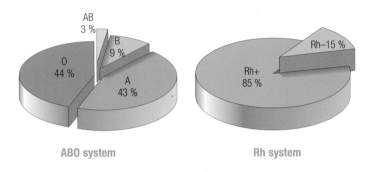

ABO system

Rh system

## Blood groups

Blood is classified depending on the presence or absence of certain antigens on the surface of the red blood cells. The presence of antigens is determined genetically and is governed by the laws of heredity. This classification determines the degree of blood compatibility, in other words, the ability to transfuse the blood of one person into another person without any negative reaction. If one transfuses the type of incompatible blood of one person with a specific blood to a person with another, it is possible that the red blood cells of the donor introduced into the blood circulation of the receiver might be attacked and destroyed by antibodies. In some cases, an incompatibility reaction can be light, but, in other cases, can result in death. Numerous antigens have been identified on the surface of the red blood cells. The most important ones, those that are usually considered at the moment transfusions are made, are those in the ABO system and the Rh factor.

## The ABO system

This system is based on the existence of two antigens on the surface of the red blood cells, called AB. Based on the presence or absence of one or both antigens, four blood types have been identified. Type A cells express only antigen A. Type B cells express only antigen B. Type AB cells express both antigens A and B. Type O cells express neither antigen A nor B.

The absence of certain antigens on the surface of the red blood cells results in the presence of specific antibodies in the plasma. These antibodies are responsible for incompatibility reactions. Type A blood contains anti-B antibodies, type B blood contains anti-A antibodies, and type O blood contains both anti-A and anti-B antibodies. Type AB blood does not contain antibodies against either A or B.

| BLOOD TYPE | SURFACE ANTIGEN | ANTIBODY |
|---|---|---|
| A | A | Y anti-B |
| B | B | Y anti-A |
| AB | A B | |
| O | | anti-A anti-B |

## ABO system transfusion compatibility

| Transfusion reaction | DONOR A | DONOR B | DONOR AB | DONOR O Universal donor |
|---|---|---|---|---|
| Antigen / Antibody | A | B | A B | |
| A anti-B | Y | | | Y |
| B anti-A | | | | |
| AB Universal recipient | | | | |
| O anti-A anti-B | Y | Y | | Y |

If a person in group A receives a blood transfusion from group B, the anti-B antibodies present in the plasma of the recipient will react against the red blood cells of the donor that contain the B antigen and will destroy them, occasionally causing deadly disorders. The same is true if blood from group A were used to transfuse a person in group B, whose plasma contains anti-A antibodies capable of destroying the red blood cells of the blood that is received. On the other hand, if a person in group AB is transfused with blood of another type, there will hardly be any problems because this person does not have either antibody. The red blood cells received will not be attacked. For this reason, one conventionally considers the person in group AB as a "universal recipient." On the other hand, a person in group O cannot receive blood from any other group because the plasma contains antibodies that would attack the transfused red blood cells. Because the red blood cells of group O do not contain antigens of any type, they can be transfused to persons in other groups without risk of being destroyed. Conventionally, people with type O are considered "universal donors."

## The Rh factor

The Rh system depends on the presence or absence of other antigens on the surface of the red blood cells, the most important one of which is called the D antigen. Around 85 percent of individuals have this antigen and are considered Rh positive (Rh+). The rest lack this antigen and are classified as Rh negative (Rh–). If blood is transfused from an Rh+ to an Rh– person, anti-Rh antibodies would be generated in the latter. If the Rh– person has a second Rh+ transfusion, he or she would destroy the blood cells that were received. Therefore, one can make Rh transfusions to Rh+ recipients but not the other way around.

## Rh factor transfusion compatibility

| | | Donor | |
|---|---|---|---|
| | Rh factor | Rh+ | Rh– |
| Recipient | Rh+ | Compatible | Compatible |
| | Rh– | Incompatible | Compatible |

## Blood transfusion equipment

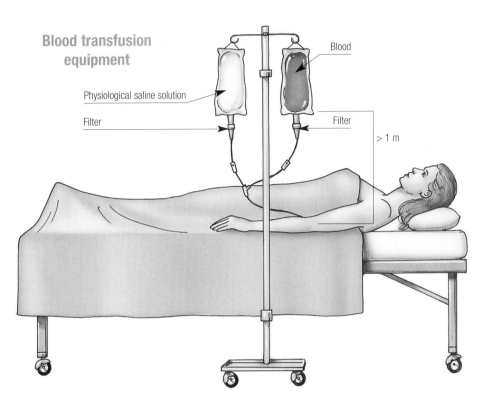

Physiological saline solution

Filter

Blood

Filter

> 1 m

The transfusion of blood or derivatives is a customary therapeutic procedure used to restore blood volume or to correct a deficit of some of its components. The procedure is simple because it involves intravenous, drop-by-drop administration that is specifically designed for this purpose. A filter prevents the passage of possible microcoagulators that might be present in the pouch.

Sometimes, blood is administered along with physiological saline serum, the only solution that is compatible with blood. Both products are evenly mixed to reduce the viscosity of the preparation. The infusion is done through a catheter or a needle with a suitable caliber, which is inserted into an accessible vein, usually in the arm. It generally takes one to two hours to transfuse one unit of blood, although complete administration never takes more than four hours.

## Abnormal transfusion reaction symptoms

- Fever and chills
- Nausea and vomiting
- Headache
- Vertigo
- Pain in neck and chest
- Difficulty breathing
- Skin eruptions
- Signs of shock

*To prevent the risk of incompatibility between the blood that is to be used and the blood of the recipient, the recipient's blood type (ABO and Rh) must be determined in order to ask the blood bank for units with suitable characteristics. Before the transfusion is actually given, to be absolutely sure that there is compatibility, one performs a safety test called a "cross match." A sample of the blood earmarked for the transfusion is mixed with a sample of the blood of the recipient and checked to make sure there is no abnormal reaction.*

# Blood coagulation

**Coagulation is a complex physiological process involving various** elements present in the blood. The objective of coagulation is of maximum importance to prevent and, above all, to promote the stoppage of hemorrhages when the integrity of the circulatory system is altered.

## Phases of stopping a hemorrhage

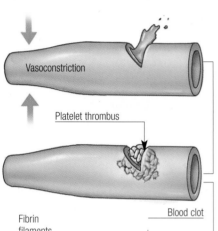

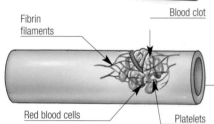

**Primary homeostasis**
If a vascular wall is injured, the blood vessel itself contracts in a reflex reaction to reduce the amount of blood that reaches the lesion. The platelets that circulate throughout the area are directed toward the lesion and adhere to each other (platelet aggregation), and a mass (the platelet thrombus) is formed. Although fragile, the thrombus plugs the breach.

**Coagulation**
A blood clot is now formed. This is a more solid and stable mass consisting of various blood elements, including red blood cells that are interlinked by abundant fibrin filaments and an insoluble protein substance that is formed from a cellular precursor, which is called fibrinogen. This is a result of the cascade action of the coagulation factors.

**Fibrinolysis**
When the damaged tissue has been repaired, the fibrin is degraded and is transformed into soluble elements. This permits the progressive dissolution of the blood clot and the normal restoration of the blood circulation in the particular area.

Under normal conditions, the walls of the blood vessels do not have any breaks in continuity that would permit the escape of blood to the outside of the vascular system. When there is a break in a blood vessel due to a trauma or a lesion that pierces its wall, the blood that circulates to the outside of the vessel tends to escape to the outside. This leads to a hemorrhage of varying magnitude depending on the size of the wound and the flow of the vessel involved. At this point, the blood coagulation mechanism goes into action in order to plug the wound.

The key point in the formation of a blood clot is the conversion of fibrinogen into fibrin, a process involving a group of elements, mostly proteins, that are usually assigned numbers. Many of these factors are always present in the plasma, but they go into action only when there is a lesion in the wall of the vessel. The mechanism is very complex because the factors involved in coagulation act in a cascade fashion. They are activated in sequence until a substance known as thrombin appears. Thrombin is responsible for converting the fibrinogen into fibrin. There are two different ways leading to the procurement of thrombin: exogenous coagulation, which is activated from substances that are secreted by the tissues of the injured blood vessel, and endogenous coagulation, which exclusively involves elements present in the plasma that go into action on contact with the injured vessel.

## Coagulation mechanism

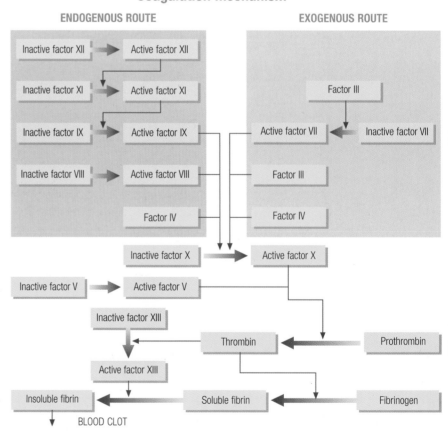

## Hemophilia

**H**emophilia is an inherited blood coagulation disorder. It is characterized by a notable tendency toward hemorrhages. It is due to a deficit or anomaly in coagulation factor VIII (hemophilia A) or factor IX (hemophilia B). When the blood coagulation abilities of either of these two factors is insufficient or when their molecular structure is anomalous, the coagulation process cannot be properly completed. Hemorrhages are then more frequent and persistent. Transmission of the affected gene is via the X chromosome, which is why most of the persons stricken are male.

## Causes of thrombosis

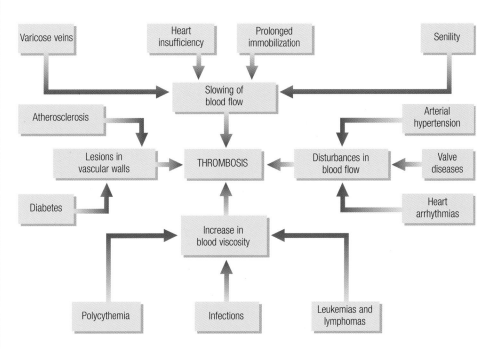

## Heredity and hemophilia

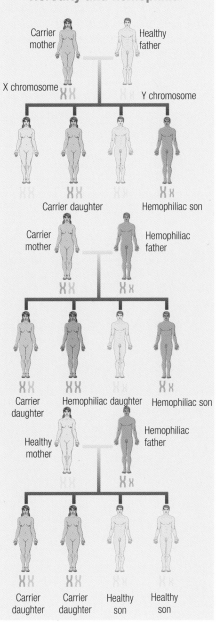

Carrier mother — Healthy father

X chromosome — Y chromosome

Carrier daughter — Hemophiliac son

Carrier mother — Hemophiliac father

Carrier daughter — Hemophiliac daughter — Hemophiliac son

Healthy mother — Hemophiliac father

Carrier daughter — Carrier daughter — Healthy son — Healthy son

## Formation of an embolus

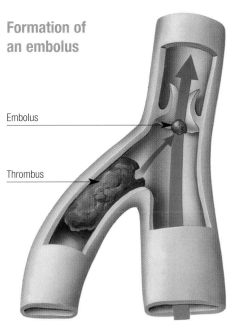

Embolus

Thrombus

*When a fragment of a thrombus breaks away, an embolus is created. It is swept away by blood flow and can be caught in a vessel with a smaller diameter, thus obstructing circulation.*

## Location of most frequent arterial thrombosis cases

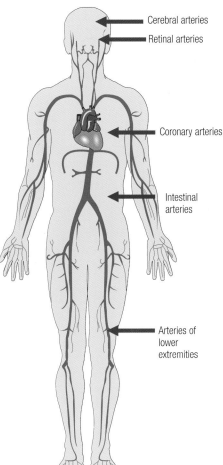

Cerebral arteries

Retinal arteries

Coronary arteries

Intestinal arteries

Arteries of lower extremities

## Thrombosis

The term "thrombosis" is used to describe the formation of thrombi or abnormal blood clots that do not stop a hemorrhage but that, on the contrary, obstruct the flow from the blood vessel and can cause an embolism. The causes are varied, and the consequences are more or less grave depending on the location. In the case of venous thrombosis, also called thrombophlebitis, the most serious complication is the separation of a fragment of the clot, which is then called the embolus. It can become stuck in an artery and thus interrupt the irrigation of the area. In the case of arterial thrombosis, the obstruction causes ischemia or an oxygen deficit in the region that is irrigated by the stricken vessel. In the most serious cases, an infarction or a necrosis of the tissue that is now deprived of oxygen and nutrients occurs.

# Blood disorders

decline in the hemoglobin in the red blood cells and is one of the most common disorders. The other is leukemia—a grave cancerous disease of the white blood cells.

## Normal levels of red blood cells and hemoglobin according to age and sex

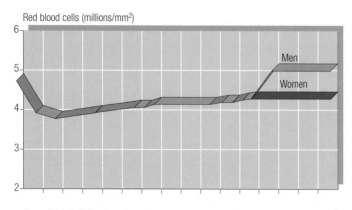

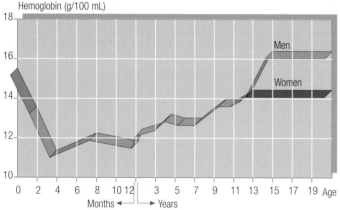

## Anemia

Anemia is a decline in the blood's hemoglobin levels. The concentration varies according to age, sex, and other circumstances. The definition of anemia depends on these parameters. Generally speaking, this disorder develops if the hemoglobin level is less than 13 g/100 mL of blood in adult men, less than 12 g/100 mL in women, and less than 11 g/100 mL in children and pregnant women. Although there is usually a relationship between the decline in hemoglobin and the decrease in red blood cells, this does not always happen. There may be no anemia even though the red blood cell count is reduced if the hemoglobin content of each one is high. On the other hand, there can be anemia even though the red blood cell count is normal if the hemoglobin content of each one is low.

The tissues do not receive all the oxygen they need to function properly when anemia is present.

Anemia can have widely varied origins. Sometimes it is due to hemorrhages that cause exaggerated or repeated loss of red cells and the hemoglobin they contain. At other times, the problem is due to a failure of the formation of hemoglobin and red cells. For example, an inherited disorder or a deficit of elements needed to produce hemoglobin, such as iron, folic acid, or vitamin B, can result in anemia. On occasion, the destruction of red blood cells is more accelerated or intense than normal (hemolytic anemia). A blood transfusion may be necessary in the most serious cases of anemia.

## Manifestations of anemia

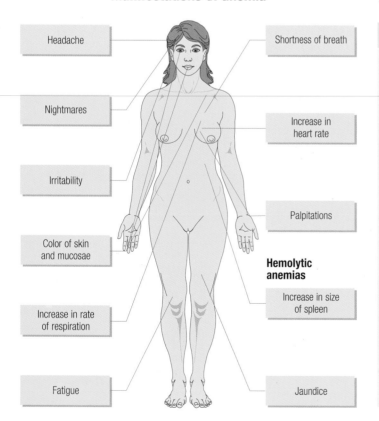

## Hemoglobin

The pigment that gives red blood cells their color is called hemoglobin. It is responsible for oxygen transport. It consists of two basic elements: a molecule called heme and a globin protein. The heme contains an iron atom that is capable of binding with oxygen ($O_2$). When exposed to a high concentration of oxygen in the lungs, the iron in each molecule of hemoglobin fixes up to four molecules of oxygen, creating **oxyhemoglobin.** As the blood circulates, the concentration of oxygen in the surrounding environment decreases while the concentration of carbon dioxide ($CO_2$), a by-product of cellular metabolism, increases. The hemoglobin releases the bound oxygen. This now-free oxygen is passed on to the tissues for use in cellular activities. At the same time, the hemoglobin binds the carbon dioxide, creating **carboxyhemoglobin.** The hemoglobin then transports the carbon dioxide to the lungs. The high concentration of oxygen in the lungs causes the hemoglobin to release the carbon dioxide for elimination through respiration. The hemoglobin then again incorporates the oxygen as part of an endless cycle that guarantees the exchange of gases between the body and the outside world.

## Types of leukemia

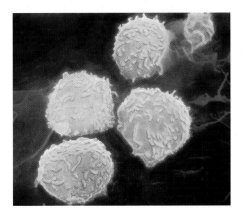

LEUKEMIA

**Origin**
Medullary | Lymphoid

**Cells that proliferate**
Granulocytes, monocytes | Lymphocytes

**Types that proliferate**
Immature | Mature | Immature | Mature

**Types of leukemia**
MYELOBLASTIC | MYELOCYTIC | LYMPHOBLASTIC | LYMPHOCYTIC

**Usual development**
ACUTE | CHRONIC | ACUTE | CHRONIC

## Leukemia

Leukemia is characterized by an exaggerated proliferation of anomalous white blood cells. There are several different forms depending on the type of white blood cells involved and how they are produced. Both polynuclear leukocytes (granulocytes) and monocytes are produced only in the bone marrow, while lymphocytes are produced in both the bone marrow and the lymphoid tissues. The anomalous white blood cells, which very often pass into the circulation in usual quantities, accumulate in the bone marrow. They invade it progressively and occupy the space set aside for the formation of normal blood cells. This process is the main reason for the consequences of this disorder. While the production of anomalous white blood cells increases, there is a decline in the formation of normal white blood cells and of the red blood cells and platelets.

The failure of some white blood cells precursors to form and the exaggerated proliferation of white blood cells is due to a genetic mutation whose underlying cause is not completely known. We do know that it is sometimes triggered by exposure to radiation or cancer-causing substances. As for its development, there are acute leukemias that develop so suddenly that they threaten life within a few weeks or months. There are also chronic leukemias that develop much more slowly and may even be asymptomatic for several years. In recent times, fortunately, the treatment of leukemia based on radiation therapy and chemotherapy as well as bone marrow transplants makes it possible to achieve excellent results, and the disease is practically cured in a large percentage of cases.

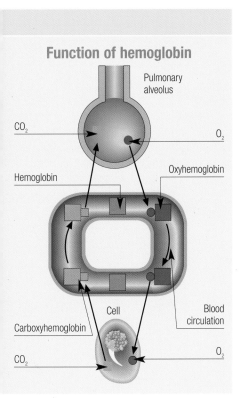

*Electron microscopic view of lymphocytes in a case of lympho-blastic leukemia.*

## Function of hemoglobin

Pulmonary alveolus

$CO_2$
$O_2$

Hemoglobin
Oxyhemoglobin

Cell
Blood circulation

Carboxyhemoglobin

$CO_2$
$O_2$

## Manifestations of leukemia

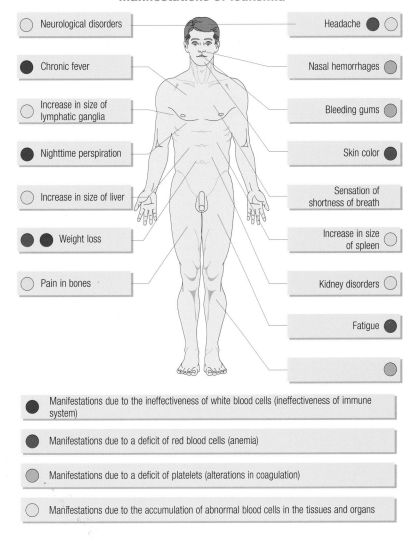

Neurological disorders | Headache
Chronic fever | Nasal hemorrhages
Increase in size of lymphatic ganglia | Bleeding gums
Nighttime perspiration | Skin color
Increase in size of liver | Sensation of shortness of breath
Weight loss | Increase in size of spleen
Pain in bones | Kidney disorders
| Fatigue

Manifestations due to the ineffectiveness of white blood cells (ineffectiveness of immune system)

Manifestations due to a deficit of red blood cells (anemia)

Manifestations due to a deficit of platelets (alterations in coagulation)

Manifestations due to the accumulation of abnormal blood cells in the tissues and organs

# Circulation

and a complex network of vessels that extends throughout the entire body. This system is responsible for continually transporting the blood that supplies the various tissues with oxygen and the nutrients they need to maintain their vital activity. At the same time, the blood picks up cellular metabolic wastes to convey them to the excretory organs.

## Cardiovascular system

**Superior vena cava**
Conveys blood poor in oxygen and containing waste from the veins in the upper part of the body to the heart.

**Pulmonary artery**
Receives blood containing carbon dioxide, pumped by the heart, and transports it to the lungs so that it can release this metabolic residue and be replenished with oxygen.

**Inferior vena cava**
Conveys blood to the heart that is poor in oxygen and contains residues from the veins in the lower part of the body.

**Veins**
Convey blood poor in oxygen and containing residues to the heart via the vena cava.

**Capillaries**
The most delicate vessels through whose fine walls exchanges take place between the blood and the tissues.

**Aorta**
The main artery of the body. It receives blood rich in oxygen, pumped from the heart, and distributes it to arteries that carry the blood to all parts of the body.

**Pulmonary veins**
Carry the blood that was oxygenated in the lungs to the heart so that the heart can pump it to the body via the aorta.

**Heart**
The pump of the circulatory system. It rhythmically pumps blood to the arteries. After passing through the entire body, the blood returns to this organ via the veins.

**Arteries**
Transport the blood pumped by the heart, rich in oxygen and nutrients, to all of the tissues of the body.

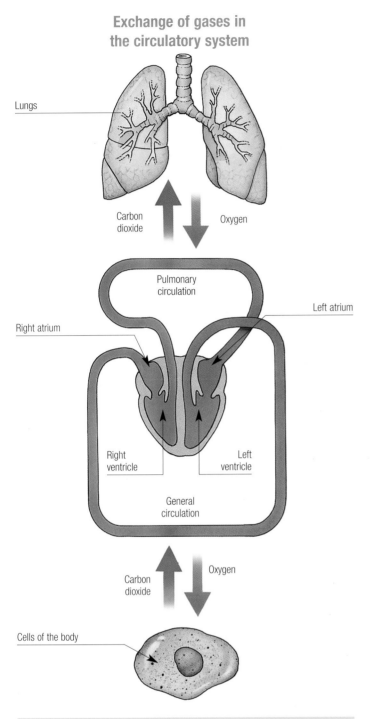

## Exchange of gases in the circulatory system

Lungs

Carbon dioxide

Oxygen

Pulmonary circulation

Left atrium

Right atrium

Right ventricle

Left ventricle

General circulation

Oxygen

Carbon dioxide

Cells of the body

## The double circulatory circuit

Although the circulatory system is a closed system, two circuits that work in a parallel way can be differentiated. This is an indispensable prerequisite for their mission. One of these circuits is called the "minor circuit" or the "small circuit" and handles pulmonary circulation. The right ventricle of the heart pumps the blood that has circulated throughout the body, is poor in oxygen, and contains carbon dioxide to the pulmonary arteries to be oxygenated. Now purified, the blood returns through the pulmonary veins through the left atrium.

The other one is called the "major circuit" or the "large circuit," which takes care of general or systemic circulation. The left ventricle of the heart propels the oxygenated blood rich in nutrients to the aortic artery so that its branches may carry it to all tissues. In the capillaries, the oxygen leaves the blood, carbon dioxide enters the blood, and the blood returns to the right atrium through the vena cava.

## Fetal circulation

**Differences between fetal circulation (left) and circulation after birth (right)**

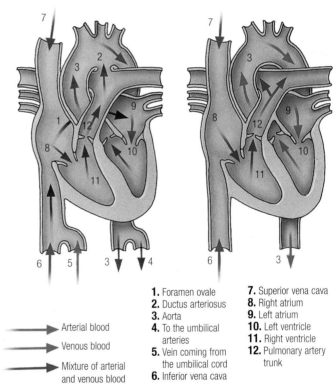

Arterial blood

Venous blood

Mixture of arterial and venous blood

**1.** Foramen ovale
**2.** Ductus arteriosus
**3.** Aorta
**4.** To the umbilical arteries
**5.** Vein coming from the umbilical cord
**6.** Inferior vena cava

**7.** Superior vena cava
**8.** Right atrium
**9.** Left atrium
**10.** Left ventricle
**11.** Right ventricle
**12.** Pulmonary artery trunk

## The vascular tree

The circulatory system is a closed circuit full of blood and made up of a system of ducts, circulatory vessels, and the pump, the heart. The heart, which is a hollow organ and consists of fatty muscular walls, is rhythmically dilated and contracted, filling up with and evacuating blood with each beat. During each contraction, the heart propels a certain volume of blood rich in oxygen to the aorta, a large artery with numerous branches, such as those that form the top of a tree. These branches repeatedly subdivide and give rise to others that are increasingly delicate, the arterioles. These finally are converted into some extremely delicate conduits, the capillaries, whose walls, made up of a single cellular layer, are so fine that they facilitate exchanges between the blood and the tissues. The capillaries are then turned into venules, which link up, forming ever-larger veins that carry blood poor in oxygen and containing metabolic waste to the heart.

Blood circulation is very different before and after birth. The fetus neither eats nor breathes, so it must get nutrients and oxygen that pass through the blood from the mother. For this purpose, the fetus has the umbilical vessels, which put its circulatory system into contact with the placenta, the organ where exchanges of substances take place between maternal blood and fetal blood. Furthermore, because there is no pulmonary circulation, the heart of the fetus has some special lines of communication that permit the passage of blood from one circulation sector to the other—an opening located in the interatrial septum wall, called the foramen ovale, and a vessel that directly links the right ventricle to the aorta, called the ductus arteriosus, which prevents the passage of blood through the lungs. After birth, as circulation through the umbilical vessels is interrupted and the infant begins to breathe, both the foramen ovale and the ductus arteriosus are closed and circulation develops through the pulmonary vessels.

# The heart

**The heart is the pump of the circulatory system. It is situated** in the chest. It continuously and rhythmically conveys blood rich in oxygen and nutritive substances through an intricate network of vessels that transport them to all of the sections of the body.

## Location of heart

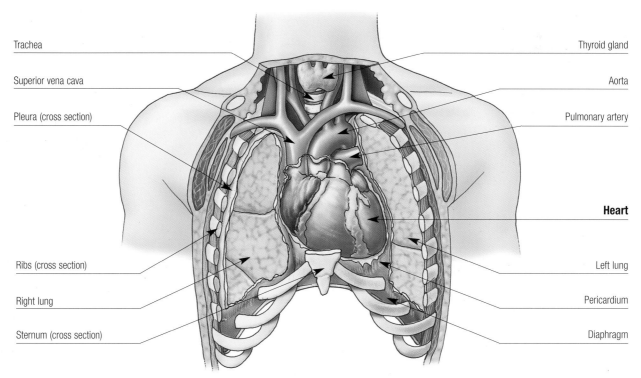

Trachea

Superior vena cava

Pleura (cross section)

Ribs (cross section)

Right lung

Sternum (cross section)

Thyroid gland

Aorta

Pulmonary artery

**Heart**

Left lung

Pericardium

Diaphragm

## Longitudinal cross section of heart

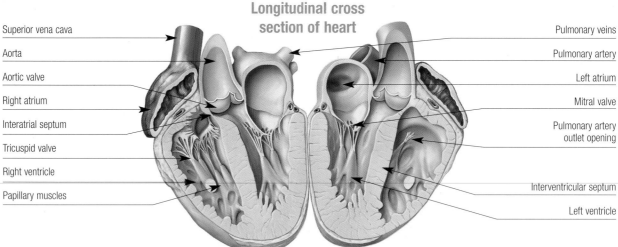

Superior vena cava

Aorta

Aortic valve

Right atrium

Interatrial septum

Tricuspid valve

Right ventricle

Papillary muscles

Pulmonary veins

Pulmonary artery

Left atrium

Mitral valve

Pulmonary artery outlet opening

Interventricular septum

Left ventricle

## External view of heart

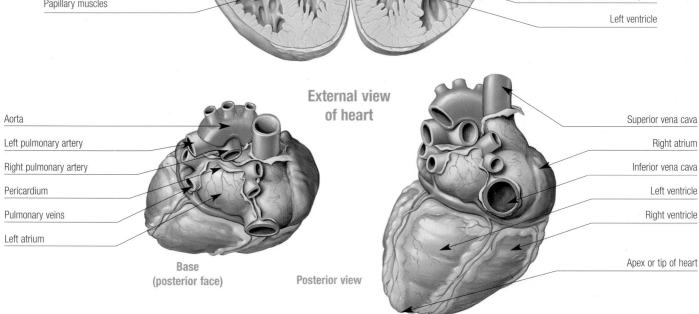

Aorta

Left pulmonary artery

Right pulmonary artery

Pericardium

Pulmonary veins

Left atrium

**Base (posterior face)**

**Posterior view**

Superior vena cava

Right atrium

Inferior vena cava

Left ventricle

Right ventricle

Apex or tip of heart

## Heart chambers

The heart is a hollow organ whose interior contains two muscular and membranous tissue walls, one vertical and the other horizontal. They create four differentiated heart compartments: the vertical wall runs across the heart from the base to the tip and divides it into two halves, a right half and a left half, which normally do not communicate with each other. The horizontal wall, on the other hand, separates the two upper chambers, called atria, from the two lower compartments, which are called ventricles. However, it has some openings that permit communication between the atrium with the ventricle on the same side.

## Innervation of heart

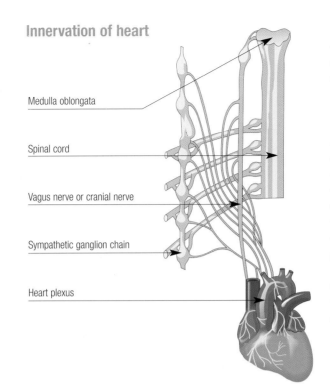

Medulla oblongata

Spinal cord

Vagus nerve or cranial nerve

Sympathetic ganglion chain

Heart plexus

*The heart is innervated by the autonomic nervous system, both by the sympathetic system through the nerves coming from the ganglion chains situated next to the thoracic spinal cord as well as by the parasympathetic system through the vagus nerve. The sympathetic system, activated by stimuli such as emotion or physical exercise, causes an increase in the heart rate. The parasympathetic system, which predominates in situations of rest and in a state of calm, slows down the heartbeat.*

## Coronary circulation

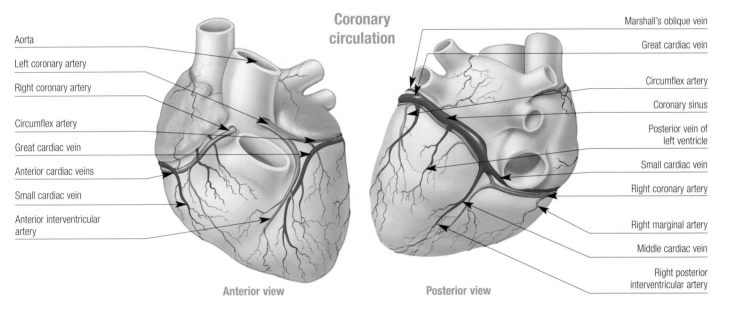

Aorta

Left coronary artery

Right coronary artery

Circumflex artery

Great cardiac vein

Anterior cardiac veins

Small cardiac vein

Anterior interventricular artery

Marshall's oblique vein

Great cardiac vein

Circumflex artery

Coronary sinus

Posterior vein of left ventricle

Small cardiac vein

Right coronary artery

Right marginal artery

Middle cardiac vein

Right posterior interventricular artery

**Anterior view**          **Posterior view**

## X-ray of heart

*The use of X rays to study the heart is extremely beneficial because it helps to determine the specific location, shape, and size of the organ and the major vessels. Although one cannot really "see" the heart in the X ray, the physician can interpret the images of this particular organ and of the major vessels, deciphering what is known as the cardiac silhouette, where each area corresponds to a specific anatomical part.*

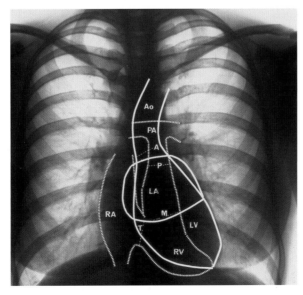

## Manifestations of heart diseases

- Palpitations
- Shortness of breath (dyspnea)
- Chest pain
- Blue coloration of skin and mucosa (cyanosis)
- Retention of fluid in tissues (edemas)

**Ao:** aorta
**PA:** pulmonary artery
**P:** pulmonary artery trunk
**LA:** left atrium
**RA:** right atrium
**LV:** left ventricle
**RV:** right ventricle
**M:** myocardium

# Heartbeat

**The heart chamber's atria and ventricles dilate and contract** rhythmically to fill up with blood and then expel their contents. This establishes a cycle that represents the pumping function of the heart and that ensures blood circulation. Electrical stimuli that are generated in the organ itself and that determine the synchronized movements of the cardiac muscle control blood circulation.

## Electrical conduction system of the heart

Internal tracts

Sinoatrial node

Atrioventricular node (AV node)

Atrioventricular bundle (bundle of His)

Right crus

Purkinje's fibers

Left crus

The heart begins to beat long before birth and functions continuously until death. Throughout an average lifetime, the heart contracts and dilates 2,500 million times.

The interrupted beats of the heart depend on **electrical stimuli** that occur autonomously in the organ itself and that are capable of causing the contraction of the fibers that make up the **heart muscle**, giving rise to the successive and sequential contraction of the heart muscle. These stimuli are generated **rhythmically** in specific sectors of the heart known as "nodes." The stimuli are produced sequentially throughout the organ by bundles and networks of specialized muscle fibers that constitute the **electrical conduction system**. Thanks to its influence, the nervous system can accelerate or slow down heart activity but cannot trigger it. The heart is an organ that is functionally autonomous.

## Propagation of electrical heart impulses

**A.** When we are at rest, the electrical stimuli are generated at the rate of 60–80 times per minute in the sinoatrial node located in the right atrium.

**B.** The stimuli are transmitted through the internodal tracts via the right atrium and the left atrium, causing the contraction of both chambers.

**C.** The stimuli reach the atrioventricular node, next to the opening that links the atrium and the ventricle on the right side. They travel via the ventricle through the crus and left crus of the bundle of His (atrioventricular bundle).

**D.** Stimuli are propagated by the right ventricle. Finally, they move on to Pirkinje's fibers that extend through the walls of the two ventricles, causing the contraction of these chambers.

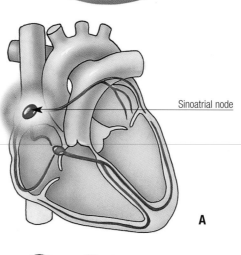

Sinoatrial node

**A**

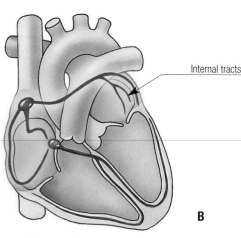

Internal tracts

**B**

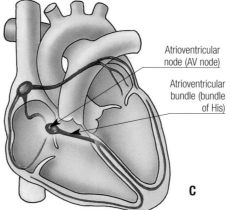

Atrioventricular node (AV node)

Atrioventricular bundle (bundle of His)

**C**

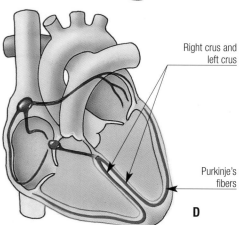

Right crus and left crus

Purkinje's fibers

**D**

The blood circulates throughout the interior of the heart in a single direction from each atrium toward the respective ventricle and from the latter toward the corresponding artery. The aorta is on the left side, and the pulmonary artery is on the right. This single-direction circulation is guaranteed by a system of valves that permits the passage of blood from one sector to the other and, in exchange, prevents its reflux. The passage of blood from the atrium to the ventricle on each side is governed by a specific valve. The right atrioventricular valve or the tricuspid valve consists of three leaflets. The left atrioventricular valve or mitral valve is made up of two main leaflets and its appearance resembles the miter covering the head of certain church dignitaries. Two more valves, the aortic valve and the pulmonary valve, are located in the openings that, respectively, link the left ventricle with the aortic artery and the right ventricle with the pulmonary artery. They permit the passage of blood flow from the ventricle only to the corresponding artery.

## Heart valves

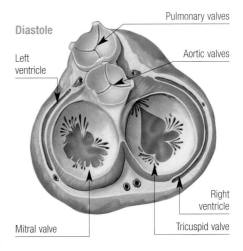

Diastole

Pulmonary valves

Aortic valves

Left ventricle

Right ventricle

Mitral valve

Tricuspid valve

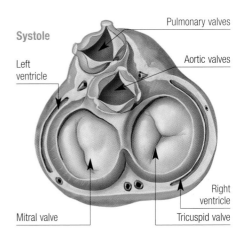

Systole

Pulmonary valves

Aortic valves

Left ventricle

Right ventricle

Mitral valve

Tricuspid valve

## Operation of heart valves

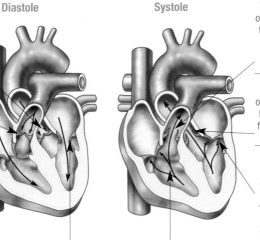

The aortic valve is closed and prevents the reflux of blood to the left ventricle.

The pulmonary valve is closed and prevents the reflux of blood to the right ventricle.

The tricuspid valve opens up and permits the passage of blood from the right atrium to the right ventricle.

The mitral valve is opened and permits the passage of blood from the left atrium to the left ventricle.

Diastole

Systole

The pulmonary valve opens up and permits the passage of blood from the right ventricle to the pulmonary artery.

The aortic valve opens up and permits the passage of blood from the left ventricle to the aorta.

The mitral valve is closed and prevents the reflux of blood to the left atrium. The tricuspid valve is closed and prevents the reflux of blood to the right atrium.

## Phases of the heart cycle

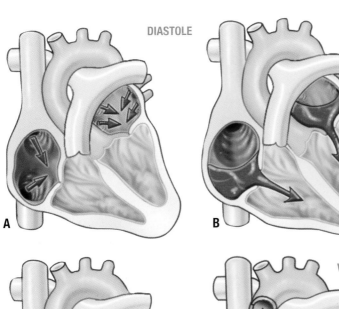

DIASTOLE

A

B

ATRIUM SYSTOLE

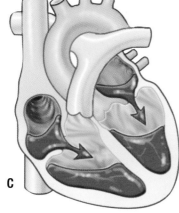

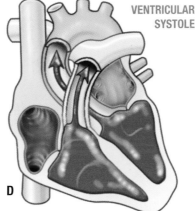

VENTRICULAR SYSTOLE

C

D

Upon each beat, the four chambers of the heart are dilated and contracted synchronously so that blood can pass from each atrium to the ventricle on its side and from the latter to the corresponding artery as part of a cycle that is constantly repeated. The dilation phase is called **diastole**, while the contraction phase is known as **systole**. On the right side, the atrium is dilated and is filled with blood from the vena cava, whereupon it contracts to convey its contents to the ventricle, which is then filled with blood. Finally, the latter is contracted to propel its contents to the pulmonary arteries. On the left side, the atrium is dilated and it is filled with blood from the pulmonary veins. It is then contracted to empty its contents into the ventricle, which is filled with blood. Finally, the latter is contracted to deliver its contents to the aortic artery.

**A.** The atria relax and are filled with blood from the veins.

**B.** The atrioventricular valves are opened and permit the passage of blood to the ventricles.

**C.** The atria contract and empty their contents into the ventricles.

**D.** The atrioventricular valves are closed, and the ventricles contract to deliver their contents into the arteries.

# Heart diseases

**At this time, diseases of the circulatory system, as a whole,** account for one of the main causes of death, especially in the developed countries, with their harmful habits (unhealthy nutrition, stress, and sedentary work). These habits, along with others, all have very negative repercussions on cardiovascular health.

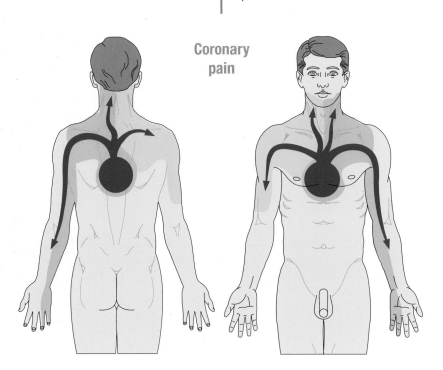

### Coronary pain

*Both angina pectoris and myocardial infarction are manifested by a pain in the center of the chest behind the sternum (retrosternal pain), which can radiate to the shoulders and even go down the arms all the way to the hands or toward the neck and the jaw or to the back.*

## Coronary disease

Coronary disease involves an alteration of the coronary arteries that are responsible for irrigating the heart and supplying the heart muscle with oxygen. The deposition of fats and other substances in the wall of the coronary arteries form plaque that narrows the opening of these vessels and prevents normal blood circulation throughout the interior or the correct irrigation of heart tissues. Plaque formation is a slow and progressive process that is silent for many years but then suddenly causes an acute myocardial infarction. There are two basic forms in which coronary disease can manifest itself: angina pectoris—severe chest pain following insufficient blood supply to the heart muscle, and myocardial infarction—the death or necrosis of a part of the heart wall due to a major and prolonged lack of blood supply.

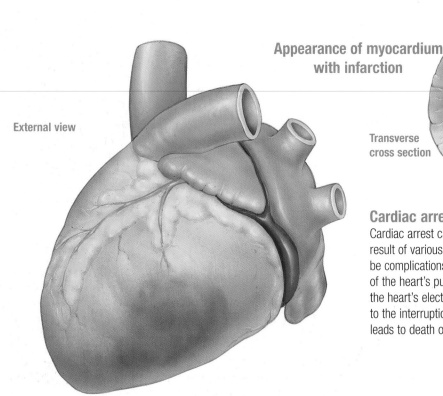

### Appearance of myocardium with infarction

**External view**

**Transverse cross section**

## Cardiac arrest

Cardiac arrest can occur suddenly in a previously healthy person as a result of various acute changes that are not always clear, or there may be complications deriving from serious and prolonged diseases. Halting of the heart's pumping action can occur basically through a disorder of the heart's electrical activity causing a cardiac arrest. This failure leads to the interruption of circulation that, on occasion, is irretrievable and leads to death or to a critical situation.

# Heart failure

Heart failure involves a weakening of the pumping action of the cardiac muscle. Two of its symptoms are shortness of breath and swelling of the legs. This disorder has many different causes, such as hypertension and coronary artery disease, depending on the part of the heart that is most widely affected and the acute or chronic development that can have multiple repercussions on the entire body. Without proper checkups and treatment, heart failure can cause serious complications and death. Today, we have resources that can boost heart function. We can therefore minimize the consequences of this disorder, provided we introduce changes in our lifestyles, and we strictly abide by the instructions given by the doctor.

## Consequences of left heart insufficiency

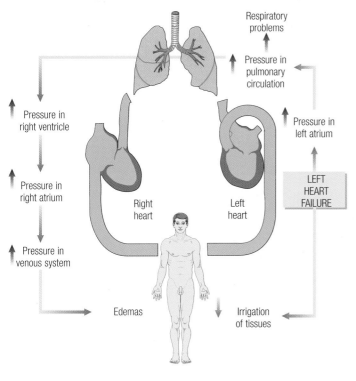

## Consequences of right heart failure

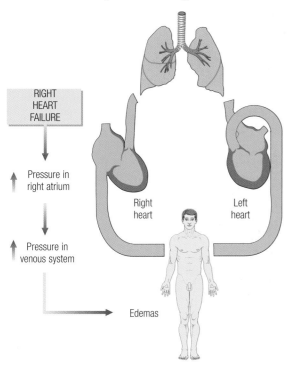

## Factors causing cardiac insufficiency

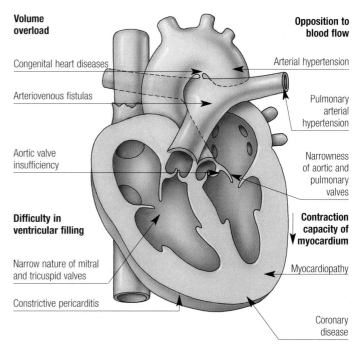

## Electrocardiogram

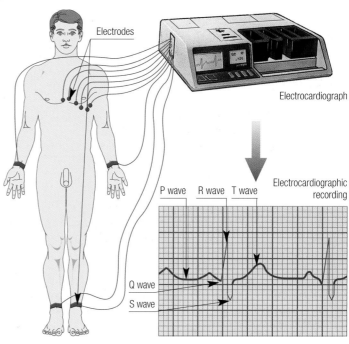

An electrocardiogram involves recording the electrical pulses that are generated rhythmically at certain points in the heart and that recur sequentially throughout the body, causing the contraction of its chambers. This test is based on detecting these electrical signals on the surface of the body, amplifying them, and translating them into some curves that are printed on paper, although it is possible to observe the tracing in a monitor. This procedure is easily performed. It is very useful in evaluating organ function. It is implemented both in response to a suspicion of certain heart diseases to arrive at a precise diagnosis and in routine health checkups.

# The arteries

**The arteries supply blood that is pumped by the heart to the** tissues of the body. The arteries branch into smaller and smaller vessels until they reach all corners of the body.

## PRINCIPAL ARTERIES OF THE BODY

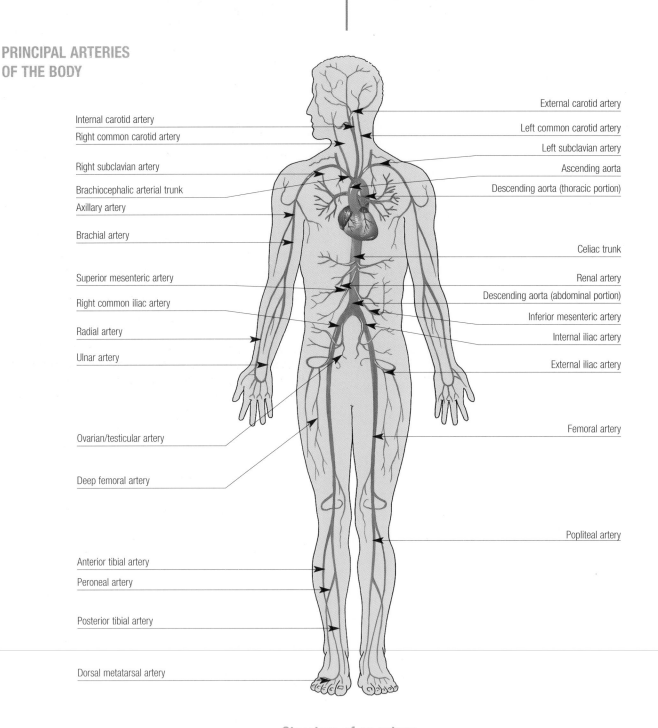

Internal carotid artery

Right common carotid artery

Right subclavian artery

Brachiocephalic arterial trunk

Axillary artery

Brachial artery

Superior mesenteric artery

Right common iliac artery

Radial artery

Ulnar artery

Ovarian/testicular artery

Deep femoral artery

Anterior tibial artery

Peroneal artery

Posterior tibial artery

Dorsal metatarsal artery

External carotid artery

Left common carotid artery

Left subclavian artery

Ascending aorta

Descending aorta (thoracic portion)

Celiac trunk

Renal artery

Descending aorta (abdominal portion)

Inferior mesenteric artery

Internal iliac artery

External iliac artery

Femoral artery

Popliteal artery

## Structure of an artery

The walls of the arteries are made up of three layers containing various tissues on which their peculiar characteristics depend:

■ **Tunica intima**, the innermost one, consists of a layer of flat epithelial cells called the endothelium, resting on a base layer that is surrounded by longitudinal elastic fibers.

■ **Tunica media**, is made up of a thin internal elastic membrane, thick muscular and transverse elastic fibers, and an elastic external layer.

■ **Tunica adventitia**, the outermost one, is a membrane made up of a connective tissue that provides strength to the tube.

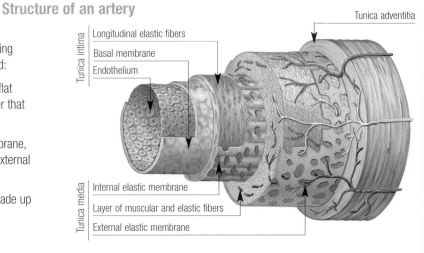

Tunica intima
Longitudinal elastic fibers
Basal membrane
Endothelium

Tunica adventitia

Tunica media
Internal elastic membrane
Layer of muscular and elastic fibers
External elastic membrane

## Arterial pulse

**W**henever the heart contracts, it forcefully propels a certain volume of blood toward the aorta, which distributes the blood via its branches. As the blood flow advances through these vessels with their elastic walls, a pulse wave is also generated along the way. That wave is caused by contraction of the left ventricle. By touching and counting the pulses felt in the superficial arteries, valuable data on heart rate, the strength of contractions of the heart, and other important parameters can be obtained.

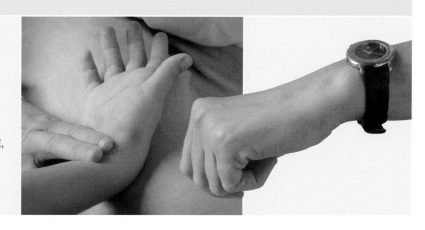

## Arterial pulse palpation points

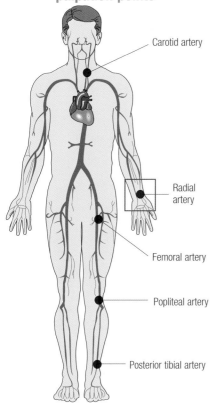

- Carotid artery
- Radial artery
- Femoral artery
- Popliteal artery
- Posterior tibial artery

## Arterial pressure

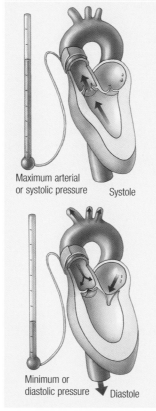

Maximum arterial or systolic pressure — Systole

Minimum or diastolic pressure — Diastole

**T**his is the force that the blood, pumped by the heart during each heartbeat, exerts against the walls of the arteries. This is a force that guarantees circulation because the blood must overcome the resistance offered by the progressive decline in the diameter of the arterial vessels. Upon each contraction, the left ventricle delivers a certain volume of blood to the aorta. The arterial branches become increasingly thinner and more elastic so that they will expand with blood and then will recover their prior diameter. As a result of this, the blood is sent to vessels with a smaller caliber, leading to the development of a practically continuous flow to the capillaries.

Arterial pressure is not uniform because it features certain oscillations during the course of the heartbeat. Therefore, when referring to blood pressure, we always consider two parameters: the maximum pressure—the systole (contraction), when the left ventricle delivers its content to the aorta, and the minimum pressure—the diastole (relaxation), when the left ventricle expands in order to be filled. In a blood pressure reading, the top number is the systolic pressure and the bottom number is the diastolic pressure. Blood pressure varies throughout the day. It also increases progressively with age, although, under normal conditions, it is always within certain limits.

## Normal limits of arterial pressure

| Age | Systolic pressure (mmHg) | Diastolic pressure (mmHg) |
|---|---|---|
| 1–3 Months | 80 | 55 |
| 4–12 Months | 90 | 65 |
| 1–4 Years | 110 | 70 |
| 5–10 Years | 120 | 75 |
| 11–15 Years | 130 | 80 |
| 16–20 Years | 135 | 85 |
| 21–30 Years | 145 | 90 |
| 31–40 Years | 150 | 90 |
| 41–50 Years | 160 | 95 |
| 51–60 Years | 165 | 95 |
| 61–70 Years | 170 | 98 |
| + 70 Years | 175 | 100 |

## Cross section of a capillary

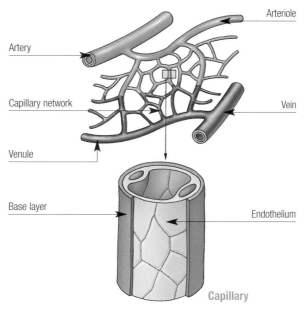

- Artery
- Capillary network
- Venule
- Base layer
- Arteriole
- Vein
- Endothelium
- **Capillary**

## Capillaries

These are continuations of the small arterioles. Their diameter is very small. They have thin walls composed of a single cell layer that permits the exchange of oxygen, nutritive substances, and waste products between the blood and the tissues they irrigate. Specifically, this is the objective of the entire cardiovascular system: to guarantee circulation in the capillaries and thus to facilitate exchanges between blood and tissues.

# The veins

**The veins are the circulatory vessels that return to the heart the** blood that comes from all parts of the body, blood that, after passing through the organic tissues, is poor in oxygen and is loaded with waste from cellular metabolism.

## Principal veins of the body

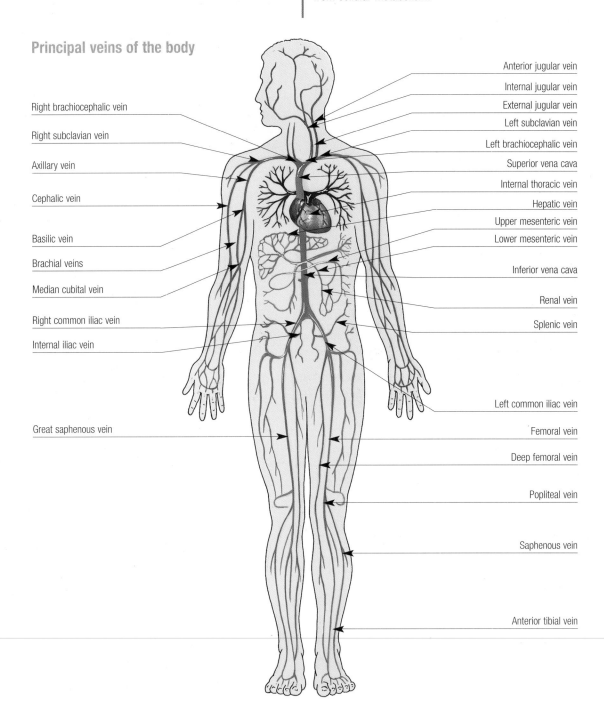

Right brachiocephalic vein
Right subclavian vein
Axillary vein
Cephalic vein
Basilic vein
Brachial veins
Median cubital vein
Right common iliac vein
Internal iliac vein
Great saphenous vein

Anterior jugular vein
Internal jugular vein
External jugular vein
Left subclavian vein
Left brachiocephalic vein
Superior vena cava
Internal thoracic vein
Hepatic vein
Upper mesenteric vein
Lower mesenteric vein
Inferior vena cava
Renal vein
Splenic vein
Left common iliac vein
Femoral vein
Deep femoral vein
Popliteal vein
Saphenous vein
Anterior tibial vein

## Structure of a vein

The walls of the veins are made up of three layers:

■ **Tunica intima**, the innermost one, is a very thin layer composed of a single layer of flat cells resting on a basal membrane of connective tissue.

■ **Tunica media**, the strongest one, is composed of elastic tissue and basically muscle tissue.

■ **Tunica adventitia** is the outermost one. It is a thin sheath of flexible connective tissue through which the adjacent layers are nourished.

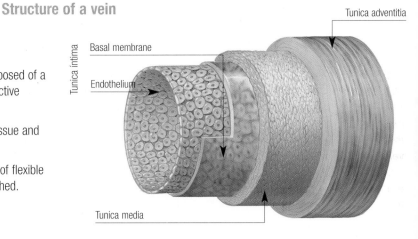

Tunica intima
Basal membrane
Endothelium
Tunica media
Tunica adventitia

## Venous system of lower limbs

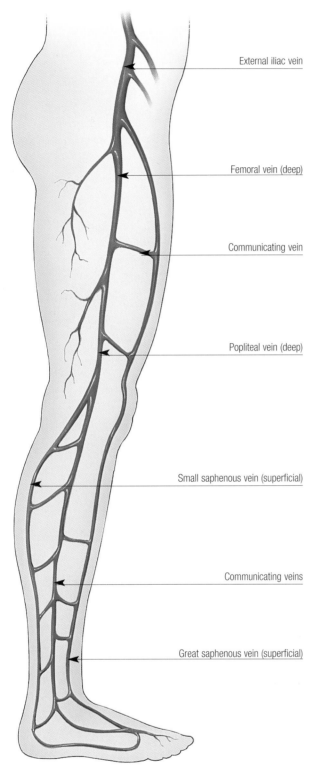

External iliac vein

Femoral vein (deep)

Communicating vein

Popliteal vein (deep)

Small saphenous vein (superficial)

Communicating veins

Great saphenous vein (superficial)

*In the peripheral venous system, we distinguish two types of veins. The **superficial veins** run very close to the surface of the body and may even be visible under the skin, especially in the extremities. The **deep veins** run between the muscles, usually following the line of the main arteries. In the lower limbs, there are also **communicating veins** that link both parts of this venous system and allow the blood to pass from the superficial veins to the deep veins. The deep veins have a larger caliber and are stronger so that they can pump the blood to the heart.*

## Venous circulation

**V**eins have the job of ensuring what is called "return circulation" because they are responsible for delivering blood that comes from all parts of the body to the heart. In the veins situated in the upper part of the body, the return is possible because the vein walls are dilatable and the pressure inside them is lower than the pressure in the right atrium, exerting an "aspiration" effect. The situation is different in the lower part of the body. When one is standing, the blood must circulate to the heart against the force of gravity. These vessels have a system of internal valves that allow blood to pass in only one direction, to the heart, while preventing its reflux. Furthermore, in the lower limbs, there is a "muscular pump," because contraction of the muscles between which the veins extend provides the thrust necessary for venous circulation.

## How the venous valves work

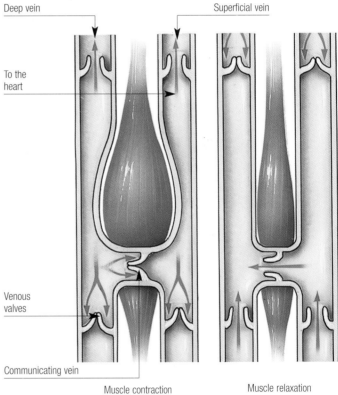

Deep vein

Superficial vein

To the heart

Venous valves

Communicating vein

Muscle contraction

Muscle relaxation

### The venous valves

Various veins in and above the lower limbs have inner valves that ensure the flow of blood in a single direction from the superficial venous vessels to the deep vessels and from the latter toward the heart. Each valve is composed of two folds of the inner wall, which have a hemispherical shape. When blood ascends, the valves are projected upward, permitting blood to flow. When the pulse is interrupted, the valve closes simply because of the weight of the blood accumulated in the concave portion of the valves. The blood cannot return downward, and in the next pulse, it will ascend along another section, always toward the heart.

# The lymphatic system

**The lymphatic system consists of an integrated network of** ducts, draining the fluid that washes the intercellular spaces and transporting it toward the circulatory system to join the blood flow. While in this network, the fluid, which is called lymph, passes through some nodular formations, or lymph nodes, that act as filters for bacteria and impurities.

## Lymphatic system

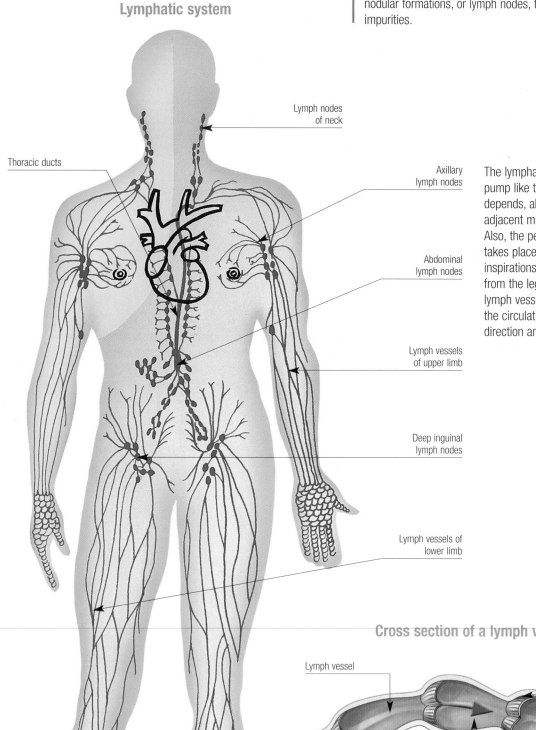

Lymph nodes
of neck

Thoracic ducts

Axillary
lymph nodes

Abdominal
lymph nodes

Lymph vessels
of upper limb

Deep inguinal
lymph nodes

Lymph vessels of
lower limb

The lymphatic system does not have a central pump like the heart, which is why its operation depends, above all, on the compression that adjacent muscles exert on the lymph vessels. Also, the periodic drop in the pressure that takes place within the chest cavity during inspirations facilitates the ascent of the lymph from the legs toward the torso. Inside the lymph vessels, a system of valves provides for the circulation of the lymph in a single direction and prevents it from flowing back.

## Cross section of a lymph vessel

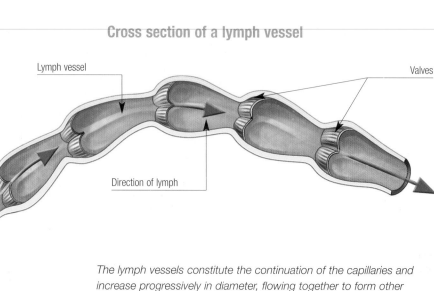

Lymph vessel

Valves

Direction of lymph

*The lymph vessels constitute the continuation of the capillaries and increase progressively in diameter, flowing together to form other lymph vessels that become even thicker.*

## Relationship between lymph and blood circulation

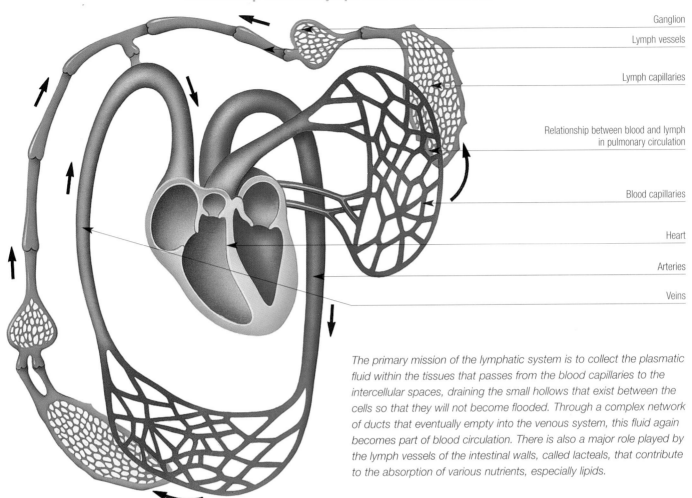

Ganglion

Lymph vessels

Lymph capillaries

Relationship between blood and lymph
in pulmonary circulation

Blood capillaries

Heart

Arteries

Veins

Relationship between blood and
lymph in abdominal circulation

The primary mission of the lymphatic system is to collect the plasmatic fluid within the tissues that passes from the blood capillaries to the intercellular spaces, draining the small hollows that exist between the cells so that they will not become flooded. Through a complex network of ducts that eventually empty into the venous system, this fluid again becomes part of blood circulation. There is also a major role played by the lymph vessels of the intestinal walls, called lacteals, that contribute to the absorption of various nutrients, especially lipids.

## Relationship between lymphatic and blood capillaries

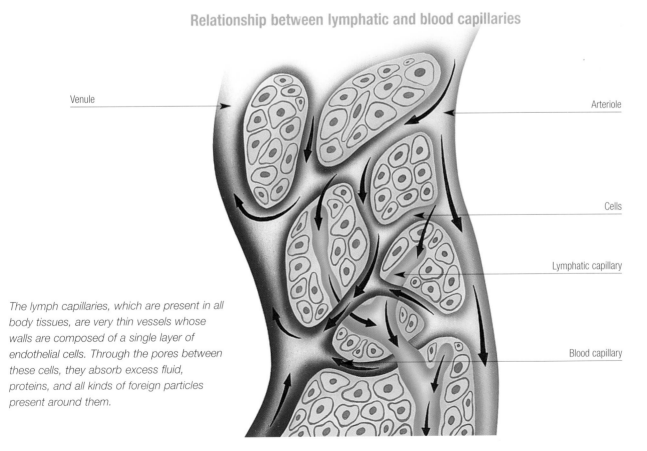

Venule

Arteriole

Cells

Lymphatic capillary

Blood capillary

The lymph capillaries, which are present in all body tissues, are very thin vessels whose walls are composed of a single layer of endothelial cells. Through the pores between these cells, they absorb excess fluid, proteins, and all kinds of foreign particles present around them.

# Respiration

**The respiratory system is the organ system responsible for** maintaining an exchange of gases with the environment and must accomplish two fundamental objectives. First, it must obtain the oxygen from the environment that our tissues use as fuel for metabolic reactions. Second, it must eliminate carbon dioxide generated as a metabolic waste because accumulated carbon dioxide is toxic.

## Components of the respiratory system

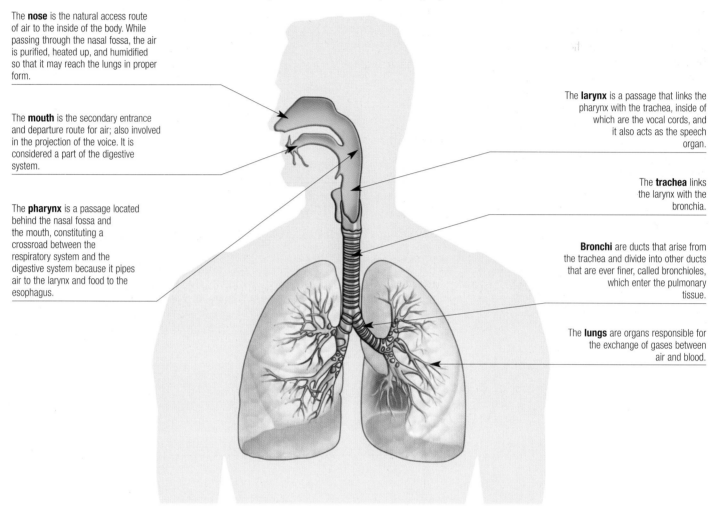

The **nose** is the natural access route of air to the inside of the body. While passing through the nasal fossa, the air is purified, heated up, and humidified so that it may reach the lungs in proper form.

The **mouth** is the secondary entrance and departure route for air; also involved in the projection of the voice. It is considered a part of the digestive system.

The **pharynx** is a passage located behind the nasal fossa and the mouth, constituting a crossroad between the respiratory system and the digestive system because it pipes air to the larynx and food to the esophagus.

The **larynx** is a passage that links the pharynx with the trachea, inside of which are the vocal cords, and it also acts as the speech organ.

The **trachea** links the larynx with the bronchia.

**Bronchi** are ducts that arise from the trachea and divide into other ducts that are ever finer, called bronchioles, which enter the pulmonary tissue.

The **lungs** are organs responsible for the exchange of gases between air and blood.

## Thoracic cage (front view)

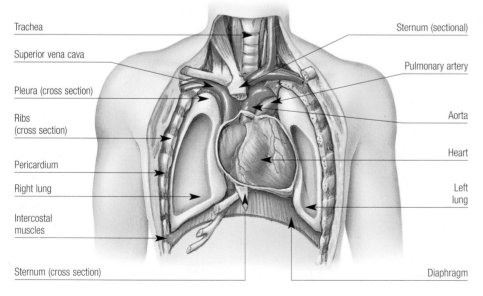

Trachea

Superior vena cava

Pleura (cross section)

Ribs (cross section)

Pericardium

Right lung

Intercostal muscles

Sternum (cross section)

Sternum (sectional)

Pulmonary artery

Aorta

Heart

Left lung

Diaphragm

The **thorax** is the part of the body between the neck and the abdomen. It contains different structures linked to various organ systems. The thoracic cavity shelters the heart and the major vessels. It also contains the major portion of the organs of the respiratory system. The chest wall is responsible for protecting viscera as delicate as the heart and the lungs in addition to very actively participating in respiration.

The diaphragm is a flattened muscle that separates the chest cavity from the abdominal cavity and is actively involved in respiration. Its convex part curves upward toward the chest cavity, and its concave part is pointed downward toward the abdominal cavity.

The diaphragm has some special openings that permit the passage of various anatomical elements from the chest cavity to the abdominal cavity. The esophagus penetrates into the abdomen through the **esophageal hiatus** to extend all the way to the stomach. The aortic artery and the vena cava pass through other openings in the diaphragm. The central part of the diaphragm or **phrenic center** is made up of hard and resistant tendinous tissue. When the strong diaphragmatic muscle contracts and flattens out, the phrenic center is pulled downward and the chest cavity is expanded.

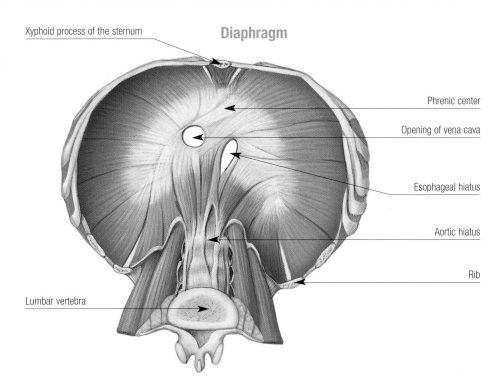

**Diaphragm**

Xyphoid process of the sternum

Phrenic center

Opening of vena cava

Esophageal hiatus

Aortic hiatus

Rib

Lumbar vertebra

## Respiration system

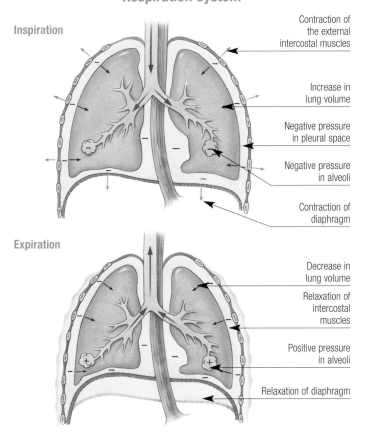

**Inspiration**

Contraction of the external intercostal muscles

Increase in lung volume

Negative pressure in pleural space

Negative pressure in alveoli

Contraction of diaphragm

**Expiration**

Decrease in lung volume

Relaxation of intercostal muscles

Positive pressure in alveoli

Relaxation of diaphragm

The entry and departure of air from the lungs is due to the action of the powerful respiratory muscles that, upon contracting and relaxing in a synchronized manner, alternately expand and contract the thoracic cage. The entrance of air from the outside into the lungs is called **inspiration**. It is due to the contraction of the diaphragm and the external intercostal muscles. The diaphragm is flattened and expands the entire thoracic cage, while the intercostal muscles elevate the lower ribs and increase the size of the thorax. The departure of air from the lungs, called **expiration**, is a passive mechanism because the lungs are elastic. When the inspiration muscles relax and stop pulling the thoracic cage, the lungs decrease to their normal volume and, as a result, expel air to the outside.

## Nervous control of respiration

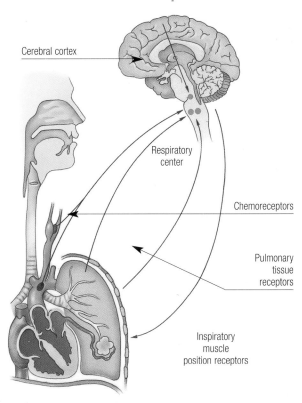

Cerebral cortex

Respiratory center

Chemoreceptors

Pulmonary tissue receptors

Inspiratory muscle position receptors

Although we can control breathing for a short time, it is typically an involuntary function. Breathing is controlled by a respiratory center located in the encephalic trunk. This respiratory center regulates the frequency and intensity of inspirations.

The respiratory center receives stimuli from both the cerebral cortex and specific receptors in various tissues and organs. Some of the receptors detect chemical changes—such as the pH, oxygen levels, and carbon dioxide levels of the blood. Others are mechanical receptors. They detect the degree of extension of the pulmonary tissue and the state of the muscles involved in breathing. The respiratory center processes all of the information from the receptors and automatically determines the optimum rhythm of respiration the body needs at each moment.

# The nose and paranasal sinuses

**The nose establishes contact between the respiratory system** and the outside world and constitutes the natural access for air to the interior of the body. The nose purifies and prepare the air so it is at the proper temperature and humidity when it reaches the lungs. The nose also acts as a sounding board when we speak, a function in which the paranasal sinuses are involved. These are cavities that are located in some bones of the cranium that are in direct communication with the nasal fossa.

## Parts of the nose

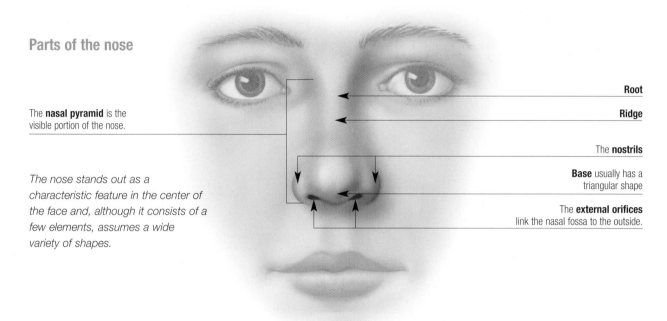

The **nasal pyramid** is the visible portion of the nose.

*The nose stands out as a characteristic feature in the center of the face and, although it consists of a few elements, assumes a wide variety of shapes.*

**Root**

**Ridge**

The **nostrils**

**Base** usually has a triangular shape

The **external orifices** link the nasal fossa to the outside.

## Anatomy of the nasal pyramid

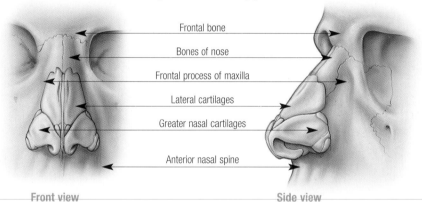

Frontal bone

Bones of nose

Frontal process of maxilla

Lateral cartilages

Greater nasal cartilages

Anterior nasal spine

Front view

Side view

*The ridge of the nasal pyramid extends from the root between the eyebrows to the tip. In many persons, it is straight, it is bent in others, and in some it is quite prominent. At the base, which has a triangular shape, are the external orifices of the nasal fossa.*

## Anatomy of the outer wall of the nasal fossa

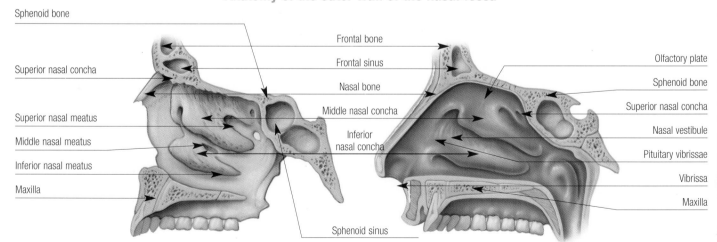

Sphenoid bone

Superior nasal concha

Superior nasal meatus

Middle nasal meatus

Inferior nasal meatus

Maxilla

Frontal bone

Frontal sinus

Nasal bone

Middle nasal concha

Inferior nasal concha

Sphenoid sinus

Olfactory plate

Sphenoid bone

Superior nasal concha

Nasal vestibule

Pituitary vibrissae

Vibrissa

Maxilla

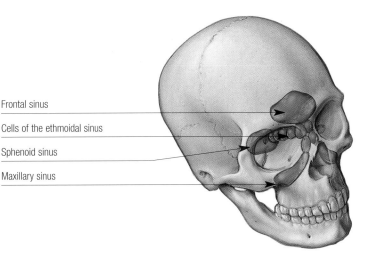

Frontal sinus

Cells of the ethmoidal sinus

Sphenoid sinus

Maxillary sinus

### The paranasal sinuses

*The paranasal sinuses are cavities that are filled with air and are located in the interior of the bones surrounding the nose. They are lined with a mucosa resembling that of the nasal fossa.*

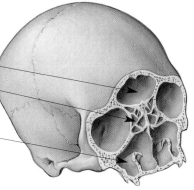

Frontal sinus

Cells of the ethmoidal sinus

Maxillary sinus

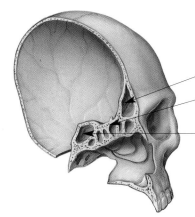

Frontal sinus

Cells of the ethmoidal sinus

Sphenoid sinus

---

## ➕ Responding to a nasal hemorrhage

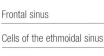

**1** Incline the torso and the head forward so the blood can flow to the outside.

**2** Breathe through the mouth, not through the nose. The air current through the nose can drag the blood clots that are forming and prevent the natural stoppage of the hemorrhage.

**3** Compress the nose, making tweezers with the index finger and thumb to exert firm and continuous pressure on the nostril of the stricken side.

**4** Maintain the pressure for five to ten minutes and then release gently to see whether the hemorrhage has stopped.

**5** If the hemorrhage persists, begin a new compression for another ten minutes and then check whether it has stopped.

**6** If the hemorrhage has not stopped after 20 minutes, continue the compression and immediately go to a health center.

---

## Causes of nasal hemorrhage

**Local disorders**

- Frailty of blood vessels
- Traumas
- Foreign bodies
- Environmental contamination
- Inflammation
- Violent sneezing
- Lesions of the nasal mucosa
- Polyps or nasosinus tumors

**Systemic disorders**

- Arterial hypertension
- Right cardiac insufficiency
- Hematological changes (hemophilia, leukemia)
- General infections (flu, measles, scarlet fever)
- Abrupt changes in pressure (mountain climbers, aviators)
- Endocrine problems and hormonal changes (puberty, pregnancy)
- Administration of anticoagulants

## Rhinitis

The most common affliction of the nose is rhinitis which involves inflammation of the mucosa that lines the nasal fossa and is characterized by nasal secretions, feelings of obstruction, and sneezing. The causes of rhinitis are highly varied, but infections and allergies stand out among them.

# The pharynx and larynx

**The pharynx is a duct consisting of muscular walls linking the** mouth and the nasal fossa to the larynx and the esophagus, which is why it is also a part of the digestive system. The larynx is a duct consisting of cartilaginous walls linking the pharynx to the trachea. It is a passage for air between the outside and the lungs in addition to being the speech organ.

## The pharynx

This is a funnel-shaped duct with a length of 5 to 6 inches (12 to 14 centimeters). Situated behind the nasal fossa and the mouth cavity, it extends into the neck and all the way up to the larynx and the esophagus. It is a part both of the respiratory system and the digestive system. Through the pharynx passes the air we breathe and also the food we ingest.

The upper pharynx is also called the nasal pharynx or **nasopharynx** and is connected along its front face with the nasal fossa. In its roof is a formation of lymphoid tissue called the pharyngeal tonsil. There is also the median pharynx or **oropharynx**, linked directly with the mouth cavity, in its anterior part. On its side faces, it has some accumulations of lymphoid tissue known as palatine tonsils, the lower pharynx, or **laryngopharynx** that connect in front with the larynx and below with the esophagus.

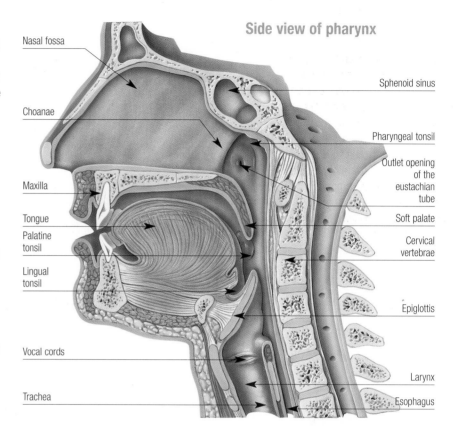

### Side view of pharynx

Nasal fossa

Choanae

Maxilla

Tongue

Palatine tonsil

Lingual tonsil

Vocal cords

Trachea

Sphenoid sinus

Pharyngeal tonsil

Outlet opening of the eustachian tube

Soft palate

Cervical vertebrae

Epiglottis

Larynx

Esophagus

Nasopharynx

Oropharynx

Laryngopharynx

### Frontal section of the pharynx (posterior view)

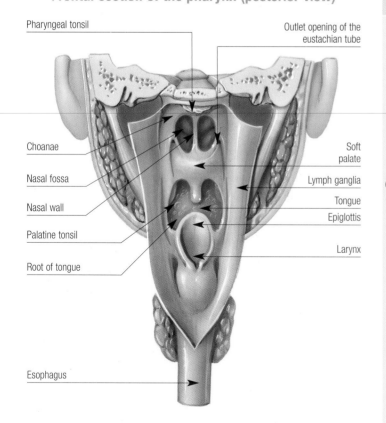

Pharyngeal tonsil

Outlet opening of the eustachian tube

Choanae

Nasal fossa

Nasal wall

Palatine tonsil

Root of tongue

Esophagus

Soft palate

Lymph ganglia

Tongue

Epiglottis

Larynx

## The pharynx while swallowing

The double function of the pharynx in the transit of air and food is possible due to the presence of the epiglottis, which is cartilage in the shape of a tennis racket situated in the upper part of the larynx that normally remains open and permits communication of air between the larynx and the outside. During swallowing, it is closed and blocks the entrance to the larynx, which forces food into the esophagus.

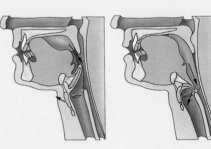
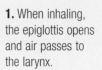
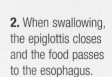
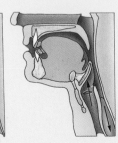

**1.** When inhaling, the epiglottis opens and air passes to the larynx.

**2.** When swallowing, the epiglottis closes and the food passes to the esophagus.

**3.** The epiglottis opens again and the air passes to the larynx.

## The larynx

This duct is in the shape of a truncated cone, composed of a series of articulated cartilages that are linked to each other by various muscles, membranes, and ligaments. Situated in the pharynx and the trachea, its dimensions vary but are small in children. The size increases during puberty, particularly in boys.

In the upper part of the larynx is the epiglottis, the cartilage whose movements either permit the free passage of air or block the duct during swallowing. Air must pass through the larynx during inspiration and expiration. The larynx has another important mission: the production of cells that create the voice. On its inside surface are two folds on each side. Some are fibrous and connect the ventricular bands or false vocal cords. Others are fibromuscular and involve the true vocal cords separated by a V-shaped cleavage known as the glottis, which is responsible for the production of sounds.

## Structure of the larynx

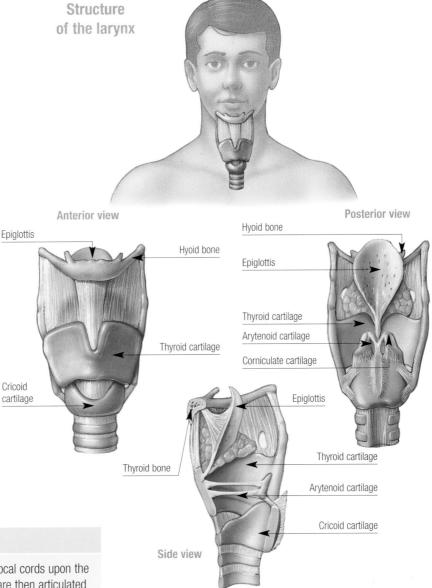

Anterior view

Epiglottis
Hyoid bone
Thyroid cartilage
Cricoid cartilage

Posterior view

Hyoid bone
Epiglottis
Thyroid cartilage
Arytenoid cartilage
Corniculate cartilage

Side view

Thyroid bone
Epiglottis
Thyroid cartilage
Arytenoid cartilage
Cricoid cartilage

## Speech

Sounds are produced by the vibration of the vocal cords upon the passage of air coming from the lungs. They are then articulated in the mouth to form the words that constitute spoken language. During inspiration and during expiration when one is not speaking, the vocal cords are relaxed and are kept folded toward the laryngeal walls so that they remain separated for a sufficient space to permit the passage of air without opposition. When one speaks, however, due to the action of the muscles that control the laryngeal cartilages during expiration, the vocal cords are tensed; they approach the median line and vibrate due to the passage of air that comes out of the lungs. This leads to the production of sounds that have different tonality depending on the degree of tension and the shape momentarily adopted by the vocal cords.

## Cross section of larynx

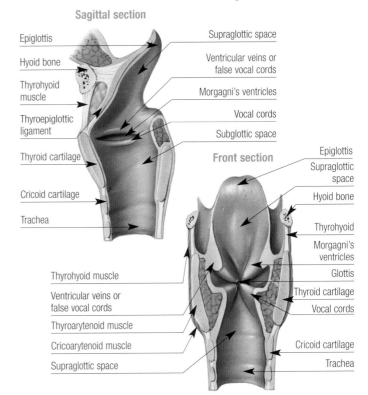

Sagittal section

Epiglottis
Hyoid bone
Thyrohyoid muscle
Thyroepiglottic ligament
Thyroid cartilage
Cricoid cartilage
Trachea

Supraglottic space
Ventricular veins or false vocal cords
Morgagni's ventricles
Vocal cords
Subglottic space

Front section

Epiglottis
Supraglottic space
Hyoid bone
Thyrohyoid
Morgagni's ventricles
Glottis
Thyroid cartilage
Vocal cords
Cricoid cartilage
Trachea

Thyrohyoid muscle
Ventricular veins or false vocal cords
Thyroarytenoid muscle
Cricoarytenoid muscle
Supraglottic space

# The trachea and the bronchi

**The trachea is a hollow organ consisting of cartilaginous walls** that link the larynx to the bronchi. These ducts, after successive branches, extend into the lungs. Together, they form the lower respiratory tracts, a necessary passage for air coming from the outside and going to the lungs and vice versa.

## Trachea

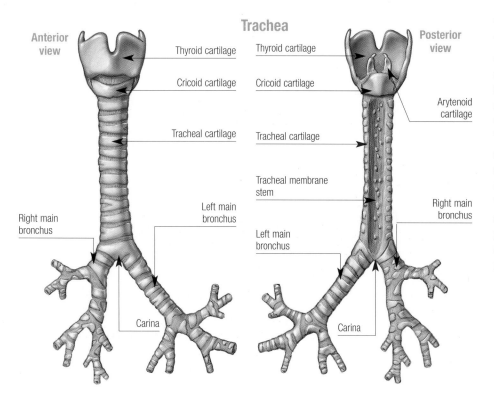

Anterior view

- Thyroid cartilage
- Cricoid cartilage
- Tracheal cartilage
- Right main bronchus
- Left main bronchus
- Carina

- Thyroid cartilage
- Cricoid cartilage
- Tracheal cartilage
- Tracheal membrane stem
- Left main bronchus
- Carina

Posterior view

- Arytenoid cartilage
- Right main bronchus

## Trachea

The trachea is a duct situated in front of the esophagus. It begins directly as a continuation of the larynx and descends through the central and anterior part of the neck to end in the chest behind the upper part of the sternum, where it divides to form the two main bronchi. The bifurcation of the duct is called the **carina**. This is a rather rigid structure since its skeleton is formed by some 15 or 20 strong cartilages in the shape of a horseshoe, open to the posterior part, but they almost completely close the circumference of the duct. The posterior part that does not cover these cartilages, the membranous part, is basically made up of connective and muscular tissue.

## The respiratory mucosa

The layer that lines the trachea and the bronchi on the inside is common to the entire respiratory tract. Essentially, it is composed of a single layer of cylindrical or cuboid cells although of different heights. The surface is covered by a series of cilia, resembling tiny eyelashes or mobile filaments. Interspersed among these cells are the calciform cells (in the shape of a calix), which are responsible for secreting mucus to the surface. The characteristics of the mucous layer make it possible to prepare the air that penetrates all the way to the lungs, humidifying it and clearing it of any impurities. The mucus forms a more or less continuous viscous film where the small, solid particles adhere that have not previously been filtered through the upper respiratory tracts. The coordinated movements of the cilia, like waves in a wheat field, move the mucus with the particles that may have adhered to it in the direction of the larynx and away from the interior of the lungs.

## Mucociliary system

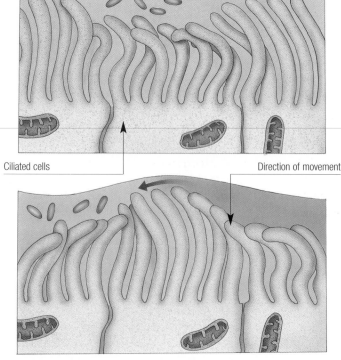

- Particles and microorganisms
- Mucus
- Cilia
- Ciliated cells
- Direction of movement

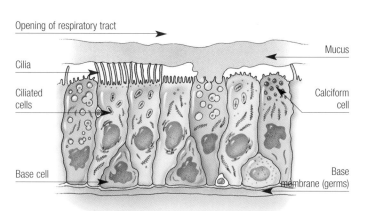

- Opening of respiratory tract
- Mucus
- Cilia
- Ciliated cells
- Calciform cell
- Base cell
- Base membrane (germs)

*The germs and tiny particles that reach the respiratory mucosa from the air coming from the outside adhere to the layer of mucus and, subsequently, are swept along in the direction of the pharynx due to the coordinated movement of the cilia.*

## Bronchial tree

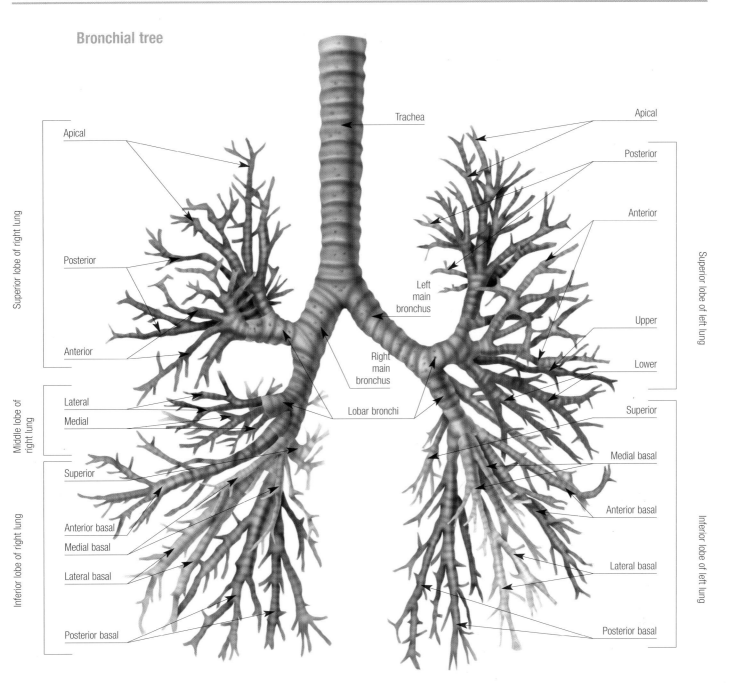

Trachea

Apical

Posterior

Anterior

Left main bronchus

Upper

Lower

Right main bronchus

Lobar bronchi

Superior

Medial basal

Anterior basal

Lateral basal

Posterior basal

Apical

Posterior

Anterior

Lateral

Medial

Superior

Anterior basal

Medial basal

Lateral basal

Posterior basal

Superior lobe of right lung

Middle lobe of right lung

Inferior lobe of right lung

Superior lobe of left lung

Inferior lobe of left lung

## The bronchi

The bronchi are a series of tubular structures with cartilaginous walls that come from the bifurcation of the trachea and, after repeated branching, extend deep into the lungs. From the trachea come two principal bronchi, each of which goes toward the corresponding lung. These bronchi are subdivided into various lobar bronchi that branch into segmentary bronchi. These continue to branch out and decrease in diameter.

A large complex of increasingly narrow branches is formed that become even thinner ducts, the bronchioles. These consist of a ciliated epithelium that does not have any cells that produce mucus. The bronchioles subdivide repeatedly, creating ever smaller ones. The terminal bronchioles are responsible for ventilating the functional unit of the lung (pulmonary alveolus). Various respiratory bronchioles empty into alveolar sacs, structures where the exchange of gases takes place between the air and the blood.

## Branches of segmentary bronchi

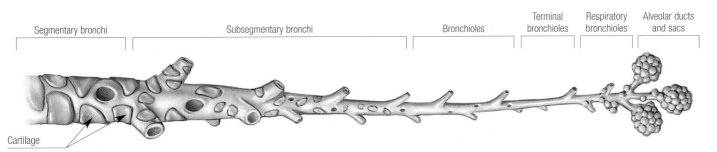

Segmentary bronchi

Subsegmentary bronchi

Bronchioles

Terminal bronchioles

Respiratory bronchioles

Alveolar ducts and sacs

Cartilage

# The lungs

**The lungs are two spongy organs housed inside the chest cavity** and linked directly to the outside through the respiratory tracts. They perform a function of vital importance: the exchange of gases between the air and the blood.

## Anatomy of lungs

**View of internal face**

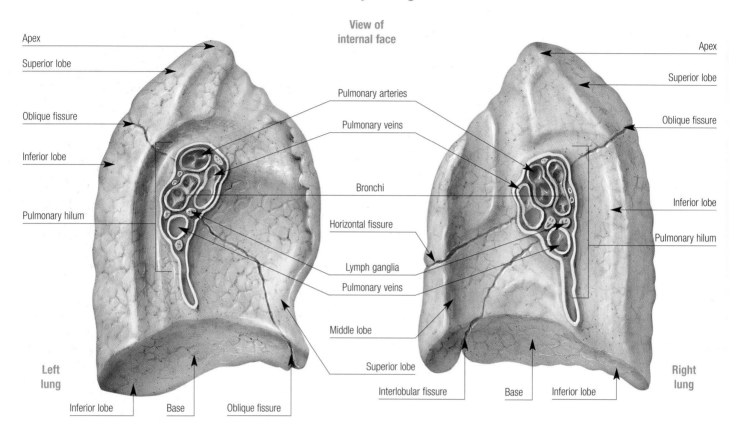

Apex
Superior lobe
Oblique fissure
Inferior lobe
Pulmonary hilum

Pulmonary arteries
Pulmonary veins
Bronchi
Horizontal fissure
Lymph ganglia
Pulmonary veins
Middle lobe
Superior lobe
Interlobular fissure

Apex
Superior lobe
Oblique fissure
Inferior lobe
Pulmonary hilum

**Left lung**

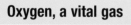

Inferior lobe     Base     Oblique fissure

Base     Inferior lobe

**Right lung**

The lungs are two voluminous organs with a semiconical shape occupying the major part of the chest cavity. Each lung has a flattened base that rests against the diaphragm, the muscle that separates the chest cavity from the abdominal cavity. The upper end, the apex, has a round shape. The lungs are grooved by deep fissures that divide them into lobes.

Each pulmonary lobe consists of various parts that are ventilated through specific bronchi. Each part, in turn, is composed of numerous secondary lobes, containing alveoli. These are tiny structures that correspond to the functional units of the lungs, because an exchange of gases between air and blood takes place in them.

## Location of lungs

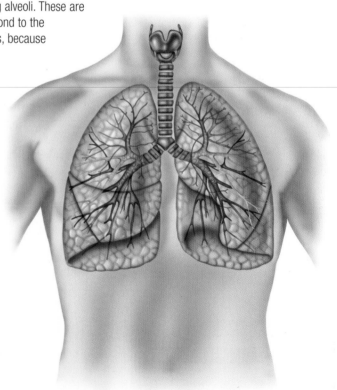

### Oxygen, a vital gas

The human body requires a constant exchange of gases with the outside. It needs to incorporate oxygen, an element that is indispensable for cellular activity and is used as fuel to obtain the energy needed for metabolic reactions. The body must also get rid of the carbon dioxide that is produced as a residue of metabolism because it is toxic. The cells require a constant supply of oxygen in order to function. For example, the neurons of the brain can subsist for only a few minutes if they do not get oxygen.

## Pulmonary alveolus

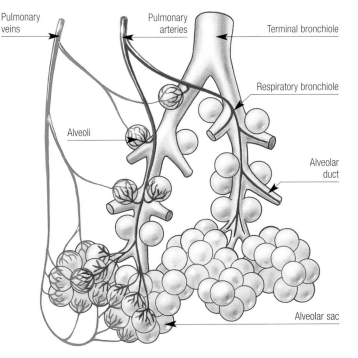

Pulmonary veins

Pulmonary arteries

Terminal bronchiole

Respiratory bronchiole

Alveoli

Alveolar duct

Alveolar sac

## Microscopic representation of alveoli and pulmonary capillaries in cross section

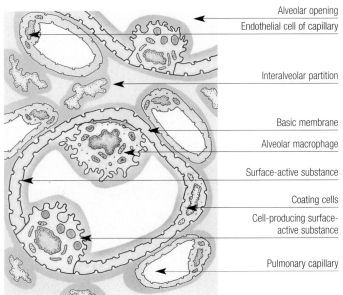

Alveolar opening

Endothelial cell of capillary

Interalveolar partition

Basic membrane

Alveolar macrophage

Surface-active substance

Coating cells

Cell-producing surface-active substance

Pulmonary capillary

*The pulmonary alveolus is the functional unit of the lungs. It is the tiny piece of tissue that is ventilated through a terminal bronchiole from the respiratory bronchioles that then give rise to the alveolar ducts. At the end of each alveolar duct is the alveoli, microscopic elastic pouches with extremely fine walls and a great deal of air. They are grouped to form a cluster or alveolar sac where gas exchange takes place.*

*The thin alveolar walls consist of a single layer of flat coating cells surrounded by a strip of supporting tissue that separates them from neighboring alveoli, the interalveolar partition. Next to the alveoli, barely separated by an extremely fine base or membrane, are the blood capillaries that "plow" through the lungs. Between the interior of one of these blood capillaries and the interior of an alveolus is a distance of less than 0.5 thousandth of a millimeter.*

## Exchange of gases

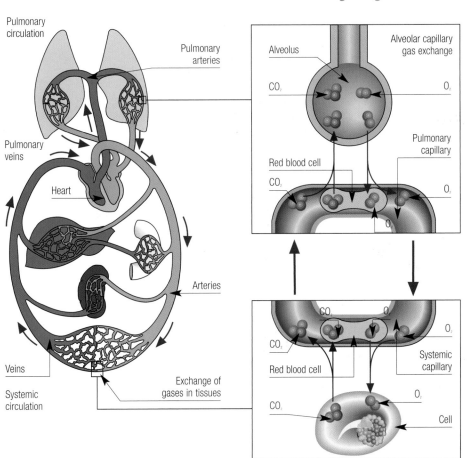

Pulmonary circulation

Pulmonary arteries

Alveolus

Alveolar capillary gas exchange

$CO_2$

$O_2$

Pulmonary veins

Pulmonary capillary

Heart

Red blood cell

$CO_2$

$O_2$

Arteries

Veins

Systemic circulation

Exchange of gases in tissues

Red blood cell

$CO_2$

$O_2$

$O_2$

Systemic capillary

$CO_2$

$O_2$

Cell

Molecules of oxygen ($O_2$) and carbon dioxide ($CO_2$) combine with the hemoglobin in red blood cells. These gases then circulate in the blood and are transported throughout the body. When the oxygen-poor blood passes through the lungs, an exchange of gases occurs in the alveoli. By simple diffusion, oxygen passes from the inhaled air to the red blood cells in the capillaries. Simultaneously, carbon dioxide diffuses from these red blood cells to the interior of the alveoli, then to be expelled from the body through exhalation.

After passing through the lungs, the oxygen-rich blood travels through the heart. The blood is then pumped to the systemic circulation, where it reaches the various body tissues. Again by simple diffusion, gases are exchanged through the capillaries. Oxygen leaves the blood and enters the body cells, while carbon dioxide leaves the body cells and enters the blood. The blood, which is now rich in carbon dioxide, travels to the lungs, where the gas exchange occurs once again.

# Digestion

**The digestive system converts the food we consume daily to its** basic nutritional components. It also ensures that the latter are absorbed by the body and transported through blood circulation to all of the corners of the body, thus supplying the materials and energy necessary to form the tissues and guarantee vital functions.

## The digestive tube

The digestive system consists of a long tube—the alimentary canal—that runs through the body from the mouth to the anus. Each digestive organ that makes up this canal has its own particular function. However, the actions of each organ are coordinated so that food is broken down into usable nutritional components, which are then absorbed by the body. The digestive system also expels wastes to the outside.

## Components of the digestive system

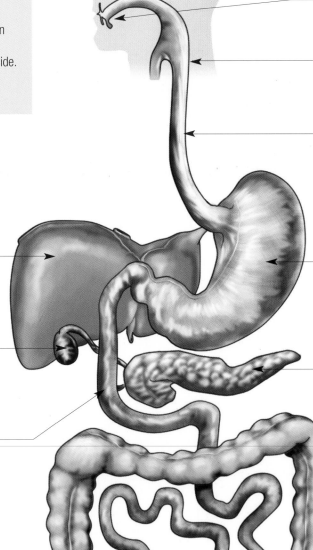

The **mouth** is the initial part of the digestive system. It is responsible for grinding the food and subjecting it to the action of the saliva in order to prepare the food for its passage through the digestive tube (alimentary canal).

The **pharynx** is the throat, located behind the mouth, that participates in swallowing.

The **esophagus** is the canal that transports the food from the pharynx to the stomach.

The **stomach** is the hollow organ that has muscular walls and stores the food, grinds it up, subjects it to the corrosive action of the stomach juices, and finally converts it into a semiliquid chyme, pushing it into the small intestine.

The **pancreas** is an organ responsible for processing a digestive juice called insulin consisting of enzymes that are indispensable for the degradation of food.

The **large intestine** is a long tract where the water of the intestinal matter is absorbed and where waste becomes fecal matter.

The **liver** is the organ that processes the bile necessary for the digestion of fats and performs various functions in the course of metabolism, especially the inactivation and elimination of toxic products.

The **gall bladder** is a hollow organ that stores the bile produced in the liver and converts it in the duodenum after meals.

The **duodenum** is the first part of the small intestine, where food is broken down due to the action of intestinal enzymes, pancreatic juices, and bile to obtain the nutritional components.

The **small intestine** is a long tube made up of three parts, the duodenum, the jejunum, and the ileum, along which the nutrients are absorbed and pass into the blood to be distributed throughout the entire body.

The **rectum** is the last part of the large intestine, where the waste from the digestive process is stored for further expulsion through defecation.

## Regions of the abdomen

The abdomen contains the organs of the digestive system. It is divided artificially along two vertical lines and two horizontal lines that subdivide the abdomen into nine parts. The upper portion consists of the left and right hypochondriac regions with the epigastric region in the center. The middle portion consists of the left and right lateral regions with the umbilical region in the center. The umbilical region contains the navel. The lower portion consists of the right and left inguinal regions with the pubic region in the middle.

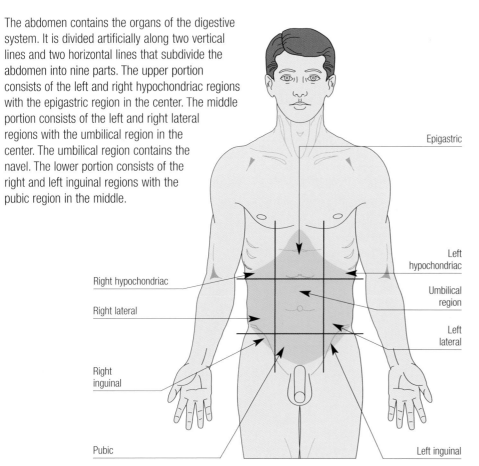

Epigastric

Left hypochondriac

Umbilical region

Left lateral

Right hypochondriac

Right lateral

Right inguinal

Pubic

Left inguinal

## Hunger and satiety

Appetite is governed by two nerve centers located in the brain, specifically the hypothalamus, the hunger center, and the satiety center. Stimulation of these centers depends on the information coming from the stomach and the receptors. When the stomach is empty, the hunger center produces a desire for eating, just as it does when one sees or smells an appetizing dish. When the stomach is full, the satiety center is stimulated, and the urge to eat disappears.

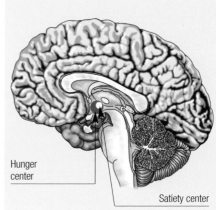

Hunger center

Satiety center

## Main intra-abdominal causes of acute abdomen

Physicians use the term "acute abdomen" to describe a grave clinical picture that is an emergency situation, frequently requiring surgery. It has a widely varied origin and is caused by a problem in the digestive system. There are numerous and diverse possible causes of this serious situation, characterized by intense and persistent pain, generally accompanied by vomiting, rigidity of the abdominal wall, and fever. This is not a disease as such but rather comes under the initial diagnosis of a very dangerous disorder that requires urgent medical examination to detect its origin and to administer the proper treatment.

LIVER AND BILE DUCTS
• Traumatic rupture
• Abscess
• Acute cholecystitis
• Acute peritonitis

SMALL INTESTINE
• Duodenal ulcer
• Obstruction
• Rupture or perforation
• Acute gastroenteritis
• Meckel's diverticulum
• Regional enteritis
• Invagination
• Intestinal tuberculosis

LARGE INTESTINE
• Ulcerative colitis
• Infectious colitis
• Intestinal volvulus
• Cancer
• Invagination
• Diverticulitis
• Rupture or perforation
• Appendicitis
• Foreign bodies

STOMACH
• Stomach ulcer
• Cancer

SPLEEN
• Infarct
• Abscess
• Rupture

PERITONEUM
• Peritonitis

INTERNAL FEMALE GENITALIA
• Rupture
• Infection
• Torsion
• Rupture of ovarian cyst
• Ectopic pregnancy
• Abscesses
• Acute salpingitis

# Mouth

**The mouth, the beginning of the digestive system, is a hollow** space, buccal or oral cavity, that consists of various structures that make possible the processing of food for digestion. The teeth tear and grind the food into fragments soaked with saliva. The tongue, the lips, and the cheeks finish the preparation of the mixture that is finally moved to the alimentary tract.

## Oral cavity

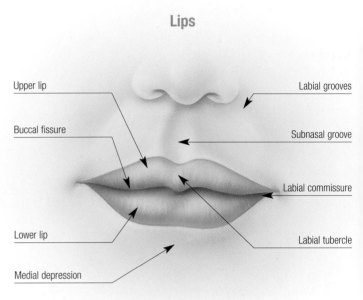

Hard palate

Soft palate

Uvula

Palatine tonsils

Palatine pillars

Lower dental arch

Lower lip

Upper lip

Upper dental arch

Palatine raphe

Buccal mucosa

Palatine pillars

Posterior wall of pharynx

Tongue

## Lips

Upper lip

Buccal fissure

Lower lip

Medial depression

Labial grooves

Subnasal groove

Labial commissure

Labial tubercle

The lips are two fleshy folds covered by skin on the outside and by mucosa inside. Their reddish color is due to extensive vascularization of the mucosa. The lips also contain numerous nerve endings that account for their great sensitivity. The lips have highly varied functions. They participate in food intake and in the production of sounds.

## Salivary glands

Various glands process the saliva, a clear and somewhat viscous alkaline fluid made up of water, mineral salts, mucin, white blood cells, and enzymes. Although numerous glands produce saliva, the main ones are distributed throughout the oral cavity. The most important ones are three pairs of salivary glands that drain their secretions into the interior of the mouth: the parotids, the submandibulars, and the sublinguals. Saliva moistens the food to facilitate mastication. It also has an antiseptic effect and contains a digestive enzyme already in the mouth that initiates the degradation of starches.

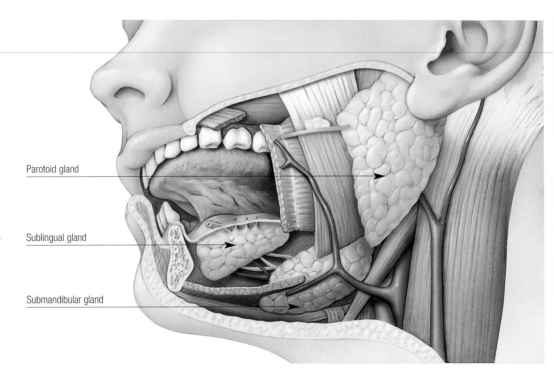

Parotoid gland

Sublingual gland

Submandibular gland

## The teeth

The teeth are hard and strong and are inserted in the maxilla and mandible. They cut (incisors), tear (canine), and grind (premolar and molar) food. Each tooth has three parts: the crown, the visible part that is over the gums; the neck, the intermediate part covered by gum; and the root, the internal part inserted in the bone. On its outer part, the crown is made of dental enamel, the hardest tissue of the body. Below is a thick layer of dentin, a softer tissue that also forms the entire root. In the center of the tooth is a cavity, the pulp, which is a soft and spongy tissue containing the blood vessels and nerves that penetrate the root of the tooth. Two sets of teeth are formed throughout a lifetime. There is a temporary set made up of 20 milk teeth. After several years, the milk teeth fall out and yield to a final set made up of 32 permanent teeth, which can never be replaced.

### Set of teeth

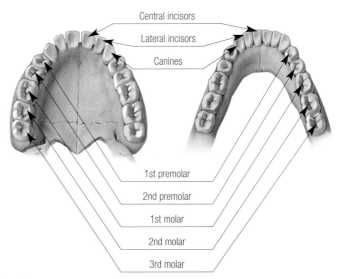

Central incisors
Lateral incisors
Canines
1st premolar
2nd premolar
1st molar
2nd molar
3rd molar

### Types of teeth

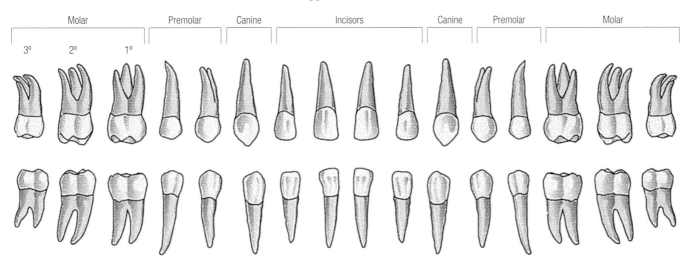

Molar — 3º 2º 1º — Premolar — Canine — Incisors — Canine — Premolar — Molar

### Parts of the tooth

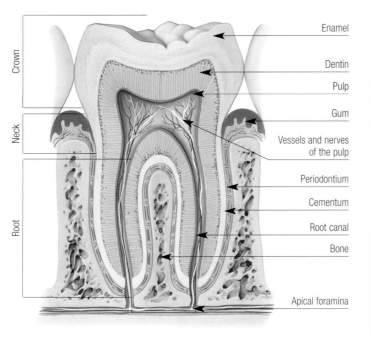

Crown — Neck — Root

Enamel
Dentin
Pulp
Gum
Vessels and nerves of the pulp
Periodontium
Cementum
Root canal
Bone
Apical foramina

### Cavities

**C**avities are very common. They involve disintegration of the hard tissues of the tooth as a result of the formation of a hole that starts on the surface and advances in depth until it causes the destruction or loss of the tooth. Among its causes are genetic factors. There is individual susceptibility to cavities as well as environmental factors, such as the action of some bacteria that are present in the mouth. These produce acids that are corrosive to the dental enamel as are sweets, which promote the formation of dental plaque and serve as fodder for bacteria.

### Development of cavities

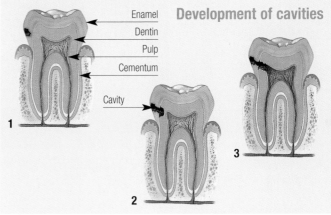

Enamel
Dentin
Pulp
Cementum
Cavity

1 2 3

# Esophagus and stomach

**The esophagus is a duct that transports the food from the throat** to the stomach, a pouch with strong muscular walls. These are lined on the inside by mucosa producing acidic substances, where the food is stored temporarily and continues through the alimentary canal.

## Esophagus (front view)

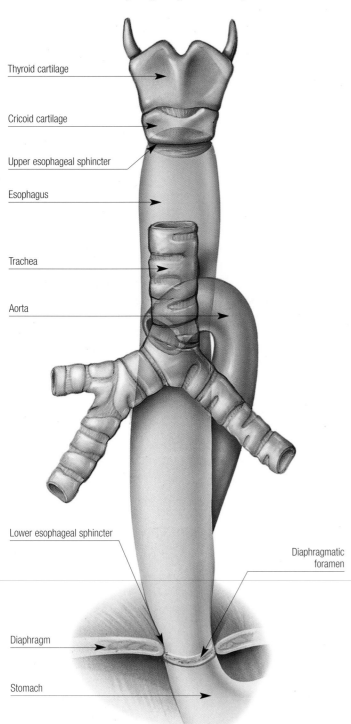

Thyroid cartilage

Cricoid cartilage

Upper esophageal sphincter

Esophagus

Trachea

Aorta

Lower esophageal sphincter

Diaphragmatic foramen

Diaphragm

Stomach

## Swallowing

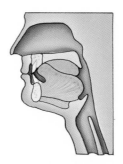
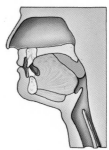
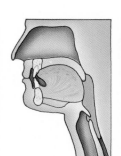

**1.** The tongue pushes the food against the palate and propels it toward the pharynx.

**2.** The velum of the palate rises to prevent the passage of the food to the nasal fossa.

**3.** The epiglottis plugs the larynx to prevent the passage of the food to the respiratory tract.

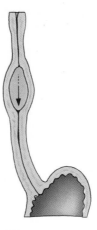

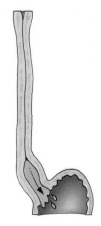

**4.** The upper esophageal sphincter opens to permit the food to enter the esophagus.

**5.** The muscles of the esophagus wall are contracted sequentially to propel the food to the stomach.

**6.** The lower esophageal sphincter opens to permit the food to enter the stomach.

The act of swallowing is a complex process in which the food passes from the mouth to the stomach through the pharynx and the esophagus. The action is initiated in a voluntary fashion by chewing the food. After that, it requires correct coordination of the movements of the various anatomical structures because some obstacles must be overcome.

The stomach is a pouch with powerful muscular walls. It starts in the pharynx, extends across the thorax behind the heart and in front of the vertebral column, and crosses the diaphragm through an opening called the esophageal hiatus. It has two cylindrical muscular formations, also called sphincters, that act as valves to keep the duct open or closed. They are at the beginning and the end of the organ.

The passage of the food from the mouth to the stomach is caused by various muscles of the pharynx and the esophagus. It is therefore possible to swallow even though we may be in a reclining position.

## Stomach (longitudinal cross section—front view)

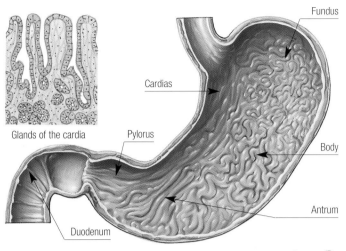

Glands of the cardia

Cardias

Pylorus

Duodenum

Fundus

Body

Antrum

The stomach is a hollow organ in the shape of a "J" that has muscular walls. The upper part is taken up by the cardia, a kind of canal that communicates with the esophagus and in the back a convex part that usually contains gases. The most voluminous part, the body, is arranged vertically. The lower part, the antrum, is horizontal and ends in a canal called the pylorus, which is a valve that remains closed until the food is prepared to continue on its way. It then opens to allow the food to pass to the duodenum. The internal surface of the stomach is lined with a mucosa that contains glands that specialize in the production of mucus and stomach juices.

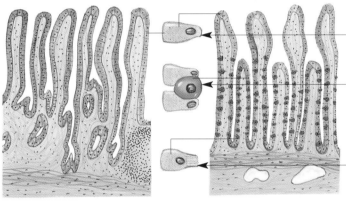

Epithelial cell (producing mucus)

Epithelial cell (producing mucus)

Principal cell (producing pepsinogen, precursor of pepsin)

Pyloric glands

Gastric or fundus glands

## Regulation of gastric secretion

The gastric juice secreted by the mucosa that lines the stomach primarily contains pepsin, an enzyme responsible for digesting proteins and releasing their amino acids components, and hydrochloric acid, a powerful corrosive needed to activate the pepsin. Although gastric juices are produced continually, this process increases when one eats. As a matter of fact, it increases simply when one thinks of eating and in response to food stimuli that one can see, smell, or taste because the nervous system stimulates the activity of stomach glands. Gastric secretion is also stimulated by a hormone called gastrin. Gastrin is released when the stomach gradually expands with food and also upon the passage of the amino acids liberated by the breakup of the proteins moving toward the small intestine.

Mental stimuli

Visual stimuli

Olfactory stimuli

Taste stimuli

Vagus nerve

Gastric juice secretion

Gastric expansion

RELEASE OF GASTRIN

Amino acids

Food

### Gastritis

**G**astritis is an inflammation of the internal mucosa of the stomach, a very common disorder with highly diverse causes, involving abdominal pain, nausea, vomiting, and, in very serious cases, even hemorrhage. The pain is usually acute, although on occasion it becomes chronic. The treatment for acute gastritis involves keeping the stomach at rest so that it may recover, followed by a very light diet for a couple of days, and avoiding hard, spicy, or slowly digested foods, such as fats.

### Causes of acute gastritis

EXOGENOUS FACTORS
- Ingestion of irritating products
- Medications (anti-inflammatory)
- Alcohol
- Improperly prepared food
- Excessive food consumption
- Heavily seasoned food
- Caustic products

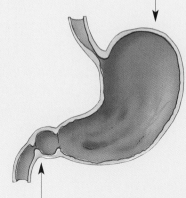

ENDOGENOUS FACTORS
- Stress (surgical procedures, burns, traumas)
- Infections (flu, hepatitis)
- Complications from other afflictions: kidney insufficiency, cirrhosis of the liver, shock

# Small Intestine

important step takes place in the digestion of food. Inside the small intestine, food is subjected to the action of enzymes from the liver, the pancreas, and the intestinal mucosa itself. The enzyme action degrades and decomposes the food into basic elements that then go through the intestinal wall and get into the circulatory system to be distributed throughout the body.

## Anatomy of the small intestine

This is a duct with a length of some 21–24 ft. (7–8 m) and a diameter of about 1 inch (2.5 cm), where three portions are differentiated, although it is continuous.

• **Duodenum:** This is the initial segment, situated at the outlet to the stomach, where the secretions of the pancreas and the bile processed by the liver are discharged.

• **Jejunum:** This is the second segment, situated in the upper region of the abdominal cavity, with a length of about 10 ft. (3 m) and with numerous curves called intestinal loops.

• **Ileum:** This is the final segment, located in the lower region of the abdominal cavity, with a length of 10–14 ft. (3–4 m), which empties into the large intestine where it discharges its intestinal content through the ileocecal valve.

### Small intestine

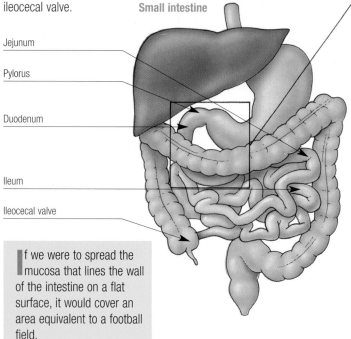

Jejunum

Pylorus

Duodenum

Ileum

Ileocecal valve

> If we were to spread the mucosa that lines the wall of the intestine on a flat surface, it would cover an area equivalent to a football field.

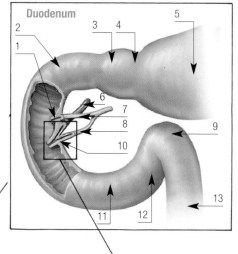

Duodenum

1. Minor duodenal papilla
2. Second portion of duodenum
3. Duodenal bulb
4. Pyloric sphincter
5. Stomach
6. Common bile duct
7. Accessory pancreatic duct
8. Main pancreatic duct
9. Arch
10. Vater's ampulla
11. Third portion of duodenum
12. Fourth portion of duodenum
13. Jejunum

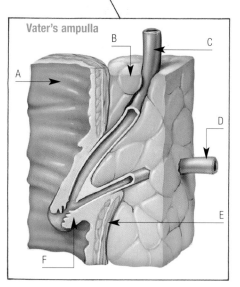

Vater's ampulla

A. Duodenal mucosa
B. Pancreas
C. Common bile duct
D. Main pancreatic duct
E. Muscle fibers (Oddi's sphincter)
F. Vater's ampulla

## The intestinal wall

The wall of the small intestine has four layers: a mucous layer, which lines the internal surface and contains a multitude of glands and secretory cells; a submucosa layer situated below the former which contains an extensive network of blood and lymphatic capillary vessels; a big muscular layer, responsible for the movements of the organs; and a serous layer that covers the duct to the outside.

The mucous layer, which lines the interior of the small intestine, has some special characteristics intended to increase the contact surface with the food and thus promote the absorption of nutrients. It has numerous tiny projections toward the opening of the organ called intestinal villi. Each of these villi has the shape of a glove finger and contains some small blood and lymphatic channels. The surface of the villi cells has an edge resembling a brush, with numerous formations resembling hairs, called microvilli.

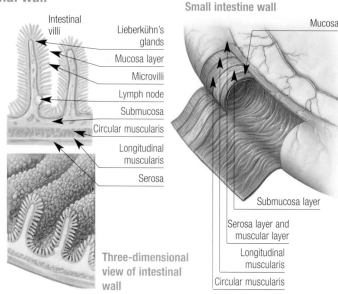

Intestinal villi

Lieberkühn's glands
Mucosa layer
Microvilli
Lymph node
Submucosa
Circular muscularis
Longitudinal muscularis
Serosa

Three-dimensional view of intestinal wall

### Small intestine wall

Mucosa

Submucosa layer

Serosa layer and muscular layer

Longitudinal muscularis

Circular muscularis

## Movements of the small intestine

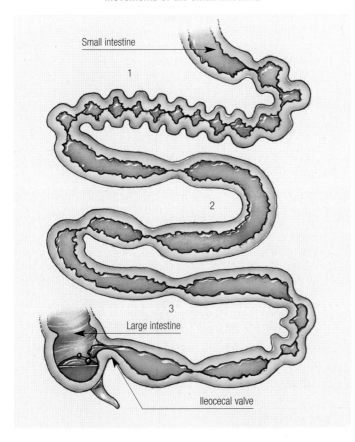

## Intestinal absorption mechanism

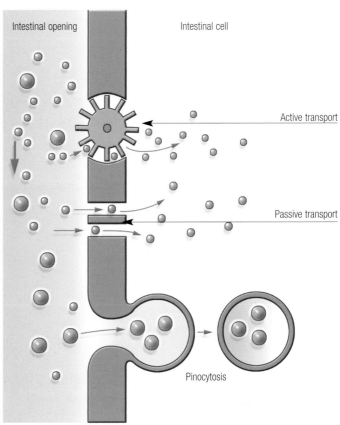

The wall of the small intestine undergoes various types of contractions that promote the mixing of the food with the digestive secretions and permit the advance of its content to the large intestine. The arrival of food coming from the stomach triggers some automatic contractions from the different intestinal segments intended to grind the food (1). There are also opposing contractions of the adjacent segments that produce a back-and-forth movement intended to mix the food with the digestive secretions (2). Finally, there are sequential contractions, the so-called peristaltic movements, that promote the advance of the food to the large intestine (3). The opening of the ileoeocal valve permits the passage of already digested food from the small intestine to the large intestine.

When food is digested by the action of the enzymes present in the lumen of the intestine, microscopic components are released that can be either absorbed or assimilated by the intestine. In other words, these components pass into the blood and lymph nodes present inside the intestinal villi. Some molecules passively penetrate into the surface cells of the gastric mucosa through tiny pores, while others are helped along by transport enzymes, and still others move on by means of a phenomenon called pinocytosis. They are "enclosed" by the membrane and are thus introduced into the interior of the cell. Once they have passed through the cells, the molecules reach the center of the villi and pass into the blood or lymphatic circulation.

## Celiac disease

Celiac disease is a chronic affliction of the small intestine caused by intolerance to gluten, a protein present in various cereals, such as wheat, barley, oats, and rye. In susceptible persons, the consumption of food containing gluten causes a series of changes in the intestinal mucosa. These make it difficult to absorb nutritive substances contained in the food. Celiac disease is manifested by diarrhea and weight loss, weakness, and other consequences of malnutrition. These problems disappear, however, when gluten is eliminated from the diet. Afflicted persons must be absolutely familiar with the makeup of the food they consume.

Food stores often offer numerous prepared foods that display this symbol, which indicates that the food item does not contain any gluten and that it is suitable for consumption by those with celiac disease.

*Permitted food*

*Prohibited food*

# Large intestine

**The large intestine makes up the final part of the alimentary** canal, the portion where food is broken down and where the nutrients are absorbed, as well as the place where the residues from digestion are stored temporarily and where waste is finally prepared for elimination in the form of fecal matter.

## Large intestine (front view)

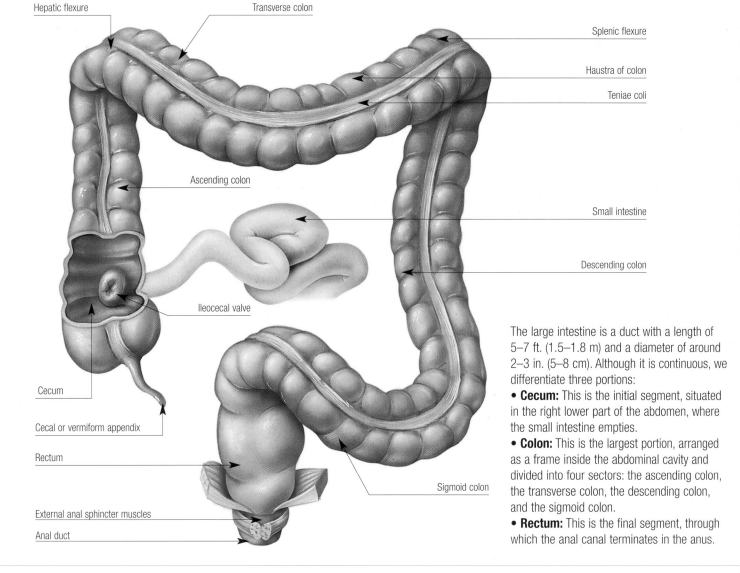

Hepatic flexure

Transverse colon

Splenic flexure

Haustra of colon

Teniae coli

Ascending colon

Small intestine

Descending colon

Ileocecal valve

Cecum

Cecal or vermiform appendix

Rectum

Sigmoid colon

External anal sphincter muscles

Anal duct

The large intestine is a duct with a length of 5–7 ft. (1.5–1.8 m) and a diameter of around 2–3 in. (5–8 cm). Although it is continuous, we differentiate three portions:
• **Cecum:** This is the initial segment, situated in the right lower part of the abdomen, where the small intestine empties.
• **Colon:** This is the largest portion, arranged as a frame inside the abdominal cavity and divided into four sectors: the ascending colon, the transverse colon, the descending colon, and the sigmoid colon.
• **Rectum:** This is the final segment, through which the anal canal terminates in the anus.

## Cross section of large intestine wall

We note four layers in the wall of the large intestine:
• The **mucosa layer** lines the entire interior of the organ. It has glands and epithelial cells. It specializes in the production of mucus and the absorption of liquids.
• The **submucosa layer** consists of slack connective tissue that contains a large network of capillary blood vessels, lymph nodes, and nerve fibers.
• The **muscle layer** is made up of two layers of muscular fibers, a circular one and a longitudinal one.
• The **serous layer**, the outermost one, is a thin tunic of fibrous elastic tissue that is an extension of the peritoneum.

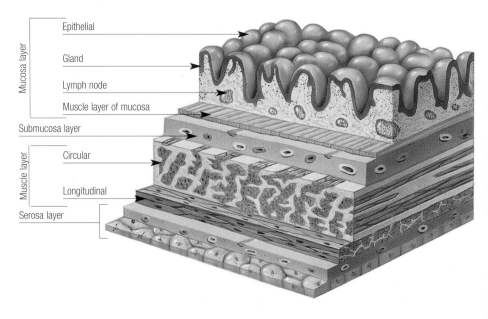

Mucosa layer

Epithelial

Gland

Lymph node

Muscle layer of mucosa

Submucosa layer

Muscle layer

Circular

Longitudinal

Serosa layer

## Anatomy of rectum and anus

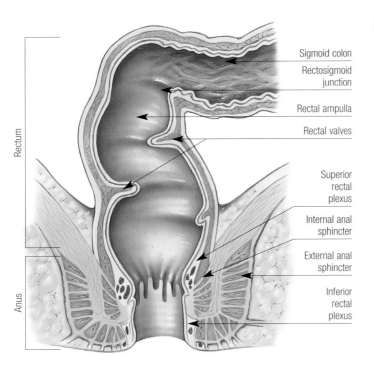

Rectum

Anus

Sigmoid colon

Rectosigmoid junction

Rectal ampulla

Rectal valves

Superior rectal plexus

Internal anal sphincter

External anal sphincter

Inferior rectal plexus

The rectum, the last portion of the large intestine, is about 6–8 in. (15–20 cm) long and has a variable diameter. It drops through the center of the pelvis and winds up in the anus, where it links to the outside. The upper part (rectal ampulla) is the most dilated portion because this is where fecal matter is stored until it is expelled. The last inch (2.5 cm) is taken up by the anal duct where we find two valves, the internal anal sphincter and the external anal sphincter, which regulate defecation.

## Formation of feces

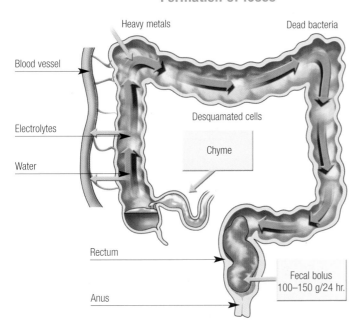

Heavy metals

Dead bacteria

Blood vessel

Desquamated cells

Chyme

Electrolytes

Water

Rectum

Anus

Fecal bolus
100–150 g/24 hr.

As it passes through the colon, the semiliquid pulp (chyme) coming from the small intestine is transformed into a compact mass called the **fecal bolus**. To the residues that are partly dried out due to the absorption of water are now added numerous dead bacteria from the intestinal flora, dried cells from the intestinal wall, and other organic waste to form the feces.

## Movements of the large intestine

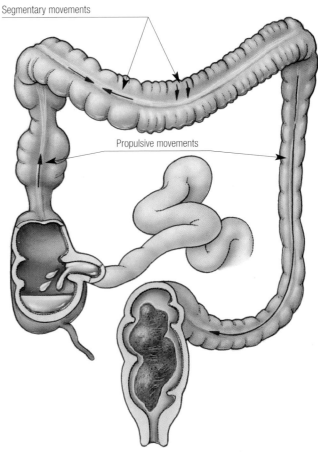

Segmentary movements

Propulsive movements

Different types of movements occur automatically and rhythmically in the large intestine. Some are segmentary, intended to mix the content and promote its contact with the walls of the organ to facilitate the absorption of water. Others are propulsive, consisting of the sequential contraction of the different segments to advance the content from the cecum to the rectum where the fecal matter accumulates. At the moment of defecation, the anal sphincter expands and the material is expelled to the outside.

One milligram of feces contains the remnants of more than 1,500,000 bacteria coming from the intestinal flora.

## Intestinal bacterial flora

The large intestine is inhabited by a large quantity of microbes that are not harmful to health. On the contrary, they are very beneficial. The germs feed on the nutrients that our body does not use, and in exchange, we gain several advantages. For example, some bacteria synthesize vitamin K and various B-complex vitamins that are absorbed by the body. The most important thing is that the presence of these bacteria under normal inoffensive conditions prevents other pathogenic microbes from settling in the intestine either because these bacteria outcompete for the food or because they produce substances that are harmful to the pathogenic microbes.

# Liver and bile ducts

**The liver is a gland that, in addition to its multiple functions** during metabolism, performs a fundamental role in digestion. It processes bile, a secretion necessary for the degradation of fats that are stored in the vesicle and routed through the bile ducts to the small intestine after each meal.

## Anatomy of the liver

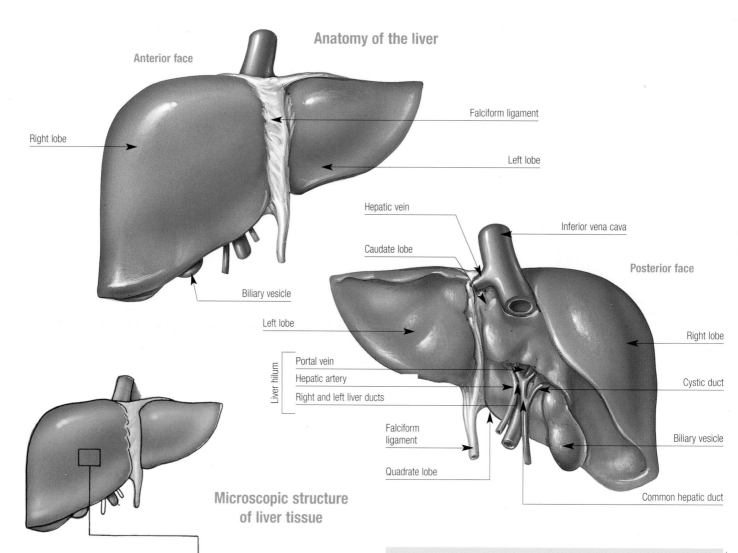

Anterior face

Right lobe

Falciform ligament

Left lobe

Hepatic vein

Inferior vena cava

Caudate lobe

Posterior face

Biliary vesicle

Left lobe

Right lobe

Liver hilum
- Portal vein
- Hepatic artery
- Right and left liver ducts

Cystic duct

Falciform ligament

Biliary vesicle

Quadrate lobe

Common hepatic duct

## Microscopic structure of liver tissue

Portal space

1   2   3        4        5        6        7

**1.** Branch of hepatic artery
**2.** Branch of portal vein
**3.** Small bile duct
**4.** Small bile canals
**5.** Liver sinusoid
**6.** Hepatocytes
**7.** Walls formed by hepatocytes

## Functions of liver

In addition to producing bile, which is the fundamental substance for digesting fats, the liver performs various functions:

▓ metabolization of the nutrients that are absorbed in the digestive tract, an indispensable step in their utilization;

▓ storage of carbohydrates in the form of glucogen, some minerals, and various vitamins;

▓ purification of numerous elements transported by the blood in the form of waste products (bilirubin, ammonia), hormones, and medications whose accumulation in the body is toxic; and

▓ synthesis of numerous substances, especially proteins and vitamins.

The liver cells, called hepatocytes, are arranged in layers that form walls around small canals that run through the entire organ. Branches of the vessels carry blood to the liver, the hepatic artery, and the portal artery from which the liver receives the substances it must process. There are also some thin, small ducts into which the hepatocytes discharge the bile they process.

## Vesicle and bile ducts

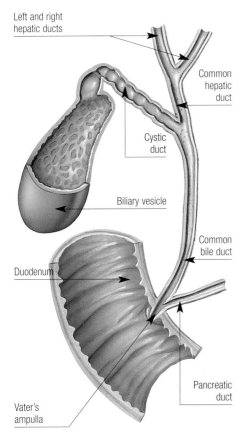

Left and right hepatic ducts

Common hepatic duct

Cystic duct

Biliary vesicle

Duodenum

Common bile duct

Pancreatic duct

Vater's ampulla

## Activity of biliary vesicle

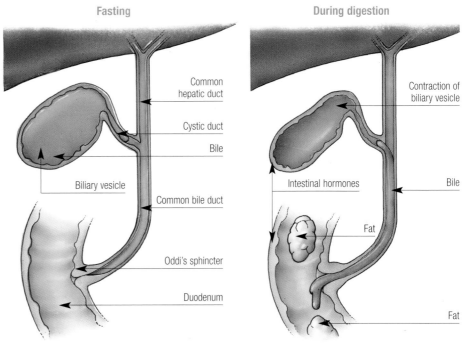

**Fasting**

Common hepatic duct

Cystic duct

Bile

Biliary vesicle

Common bile duct

Oddi's sphincter

Duodenum

**During digestion**

Contraction of biliary vesicle

Intestinal hormones

Bile

Fat

Fat

The production of bile is a constant process, but this secretion is necessary only after eating. During periods of fasting, the bile that leaves the liver through the hepatic ducts is diverted to the biliary vesicle, a hollow organ shaped like a sac, where it accumulates and is concentrated. During digestion, some hormones produced by the intestine act upon the biliary vesicle and cause it to contract and expel its contents. At the same time, a valve opens that regulates communication between the bile ducts and the intestine through which the bile is discharged into the interior of the duodenum.

## Viral hepatitis

### Mechanisms involved in spreading viral hepatitis

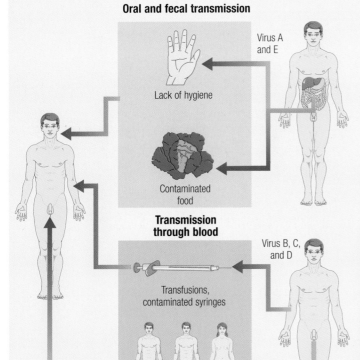

**Oral and fecal transmission**

Virus A and E

Lack of hygiene

Contaminated food

**Transmission through blood**

Virus B, C, and D

Transfusions, contaminated syringes

Sexual relations

Viral hepatitis is an infectious disease that causes inflammation of the liver and the subsequent alteration of its functions. Various viruses attack the liver, which is why we differentiate separate types of hepatitis (A, B, C, D, and E) with different forms of contagion. This affliction is manifested with some symptoms that are not at all specific, such as fatigue and loss of appetite. Later, additional signs of liver involvement appear, such as jaundice (yellowish coloration of skin and mucosae), dark urine, and light-colored feces with a claylike appearance. Sometimes manifestations are minimal. In a small percentage of cases, there is liver damage that is so abrupt and intensive that we speak of fulminant hepatitis. Manifestations of hepatitis continue for between two and six weeks and then go into remission, although development differs according to the type of hepatitis. Hepatitis A and E never develop in a chronic fashion. On the other hand, some cases of hepatitis B, C, and D tend to become chronic.

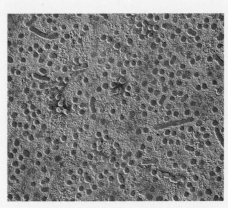

*Hepatitis B virus viewed with the electron transmission microscope.*

# Pancreas

**The pancreas is a glandular organ responsible for producing a** digestive secretion that is rich in enzymes and indispensable in the breakdown of food as well as the use of nutrients contained in the food. It is also considered an endocrine organ because it produces important hormones such as insulin to regulate the blood concentration of glucose.

### Location of pancreas

This is an elongated organ with a conical shape. It is situated transversely in the upper part of the abdomen. It has a yellow color. Its surface presents a characteristic nodular aspect that reflects its glandular structure. Although it is homogeneous, we differentiate various portions in this organ: the head, the body, and the tail. The most voluminous part, the head, is marked by the duodenum, into which it discharges its digestive secretions.

### Anatomy of pancreas

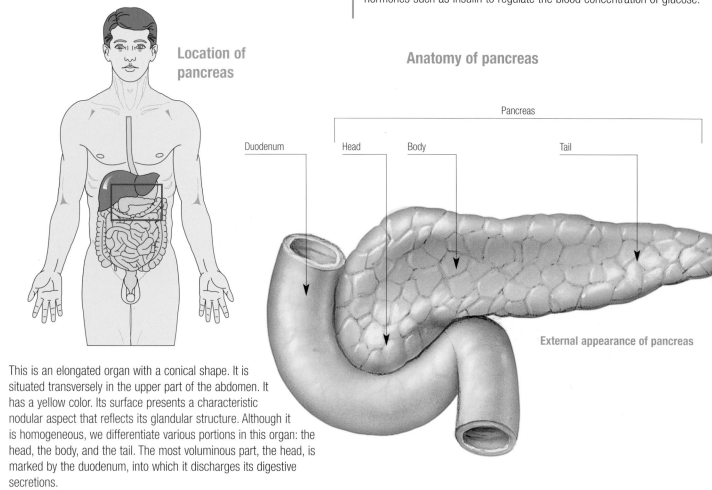

Pancreas

Duodenum    Head    Body    Tail

**External appearance of pancreas**

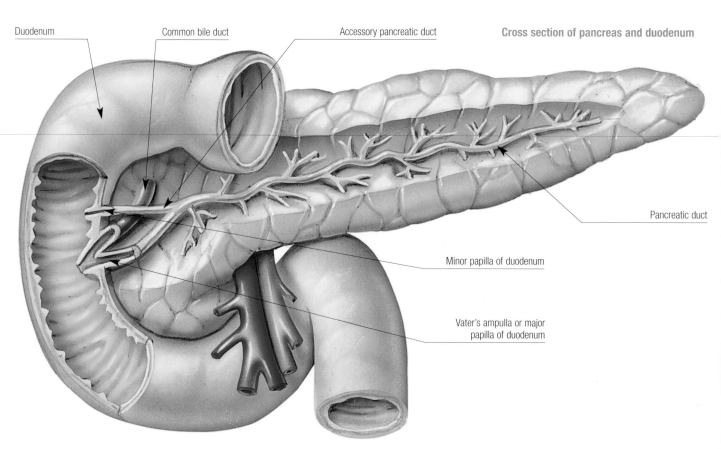

Duodenum    Common bile duct    Accessory pancreatic duct

**Cross section of pancreas and duodenum**

Pancreatic duct

Minor papilla of duodenum

Vater's ampulla or major papilla of duodenum

## Microscopic view of pancreatic tissue

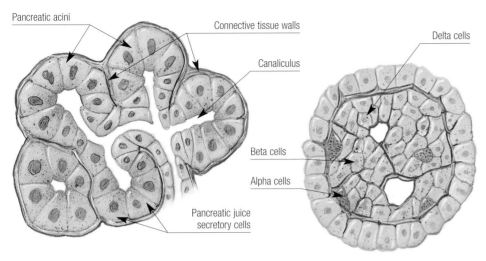

Pancreatic acini

Connective tissue walls

Canaliculus

Delta cells

Beta cells

Alpha cells

Pancreatic juice
secretory cells

Digestive pancreas

Endocrine pancreas
(Isles of Langerhans)

A multitude of glandular acini are inside the pancreas. They are the pancreatic acini consisting of a single layer of cells situated around a central opening, where they discharge the secretions they produce. Each acinus is linked to a small canaliculus that also receives the secretions of many other nearby acini. The canaliculi deliver their content to ducts that carry the pancreatic juice to the small intestine. Distributed around the entire organ surrounded by pancreatic acini are a multitude of small formations, the Isles of Langerhans. These are made up of cells that produce hormones passing directly into the blood circulation.

## Regulation of pancreatic juice production

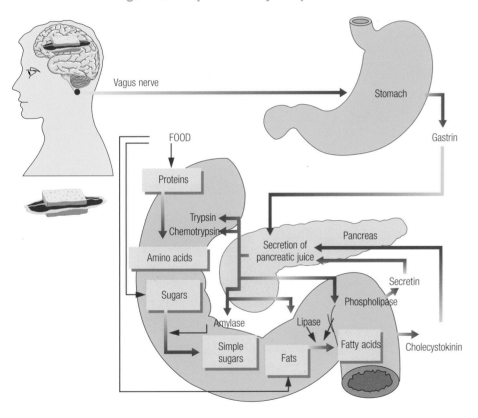

Vagus nerve

Stomach

Gastrin

FOOD

Proteins

Trypsin

Chemotrypsin

Amino acids

Sugars

Amylase

Simple sugars

Fats

Lipase

Pancreas

Secretion of pancreatic juice

Secretin

Phospholipase

Fatty acids

Cholecystokinin

The secretion of pancreatic juice is regulated by two mechanisms; one is neurological and the other hormonal. The former is linked to the autonomic nervous system. The autonomic nervous system uses the vagus nerve and the stomach processes gastrin to stimulate the production of pancreatic enzymes. These enzymes are produced in response to various stimuli—seeing, smelling, or tasting food. The secretion of pancreatic enzymes must be stimulated by hormones produced in the small intestine as it expands in anticipation of food. These hormones are called secretin and cholecystokinin.

## Pancreatitis

This is an inflammation of the pancreas, a disorder that can be either acute or chronic. It is painful. Digestive manifestations sometimes develop. They can be either minor or give rise to complications that threaten the patient's life. The most common complication is the presence of calculi in the bile ducts. These are embedded in the area that links the final part of the bile ducts to the pancreatic duct. The outflow of pancreatic juice is prevented. The fluid accumulates inside the pancreas, activating enzymes inside the organ that result in self-digestion of the pancreatic tissue. Other less-frequent causes of pancreatitis involve alcohol abuse, the actions of certain medications, infections, abdominal trauma, or a tumor in the bile ducts.

**Location of pain in the
case of acute pancreatitis**

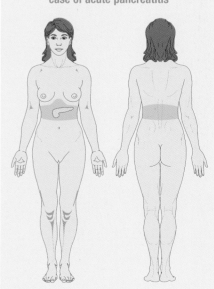
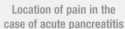

# Abdominal cavity

diaphragm, a flat muscle that separates the abdomen from the chest cavity, is the part of the torso between the chest and the lower part of the pelvis. Inside we find a good portion of the digestive system organs, although it does contain some organs belonging to the urinary system and to the reproductive system.

## Sagittal cross section of abdominal cavity

## Contents of abdominal cavity

The abdominal cavity contains organs from various systems. Digestive—stomach, small intestine, large intestine, liver, biliary vesicle, bile ducts, and pancreas. Urinary—kidneys, ureters, and urinary bladder. Reproductive—uterus, ovaries, and fallopian tubes in women; prostate gland, deep fascia of penis, and seminal vesicle in men. Circulatory—spleen. Endocrine—adrenal glands. In addition, numerous blood and lymph vessels, ligaments, and other structures fix the organs in their respective positions.

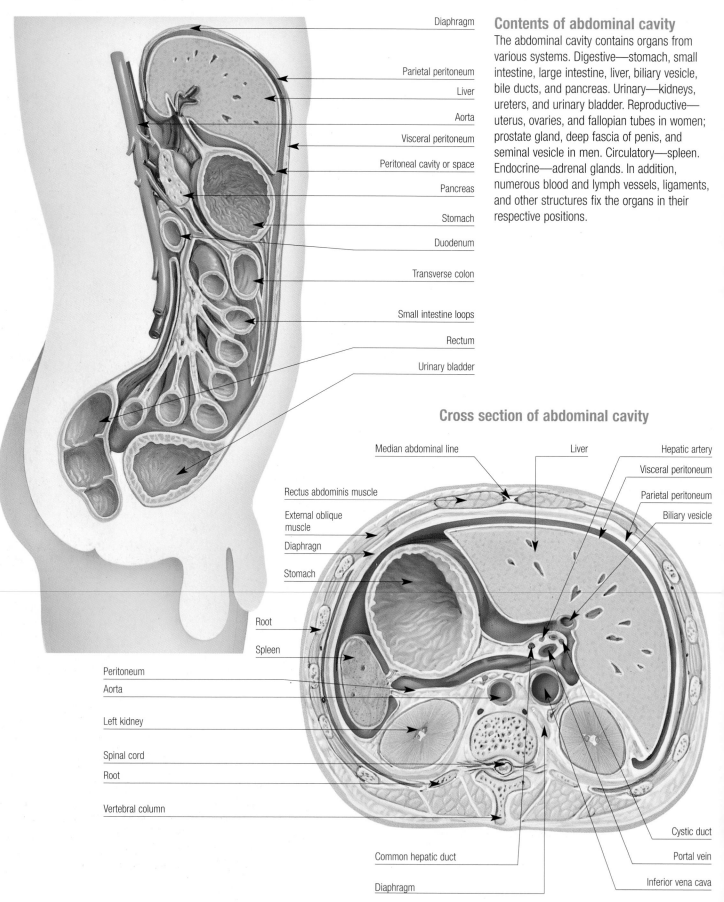

Diaphragm

Parietal peritoneum

Liver

Aorta

Visceral peritoneum

Peritoneal cavity or space

Pancreas

Stomach

Duodenum

Transverse colon

Small intestine loops

Rectum

Urinary bladder

## Cross section of abdominal cavity

Median abdominal line

Liver

Hepatic artery

Visceral peritoneum

Parietal peritoneum

Biliary vesicle

Rectus abdominis muscle

External oblique muscle

Diaphragm

Stomach

Root

Spleen

Peritoneum

Aorta

Left kidney

Spinal cord

Root

Vertebral column

Cystic duct

Portal vein

Inferior vena cava

Common hepatic duct

Diaphragm

## Abdominal hernia

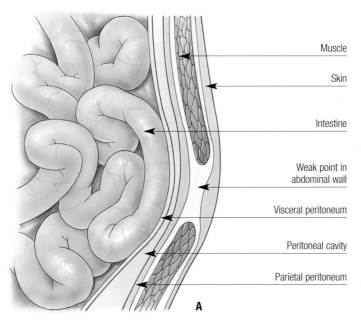

Muscle

Skin

Intestine

Weak point in abdominal wall

Visceral peritoneum

Peritoneal cavity

Parietal peritoneum

**A**

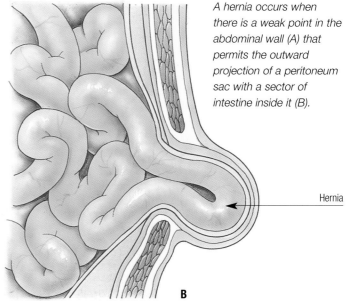

*A hernia occurs when there is a weak point in the abdominal wall (A) that permits the outward projection of a peritoneum sac with a sector of intestine inside it (B).*

Hernia

**B**

## Abdominal hernia

An abdominal hernia consists of the projection or partial protuberance of an organ, such as the small or large intestine, through a natural opening or a weak point in the abdominal wall. It looks like a painless and soft object on the body surface. It can be due to the abnormal persistence of an opening in the abdominal wall present at the time of development or due to the weakening of some point in the abdominal wall, for example, following pregnancy, obesity, repeated evacuation efforts, and others. Through this opening, propelled by the pressure from the inside of the abdomen, protrudes the peritoneum, which is a hernial sac. Sometimes there is a portion of the intestine or some other abdominal organ. Surgery is the only effective and definitive treatment of hernias.

## Types of abdominal hernia

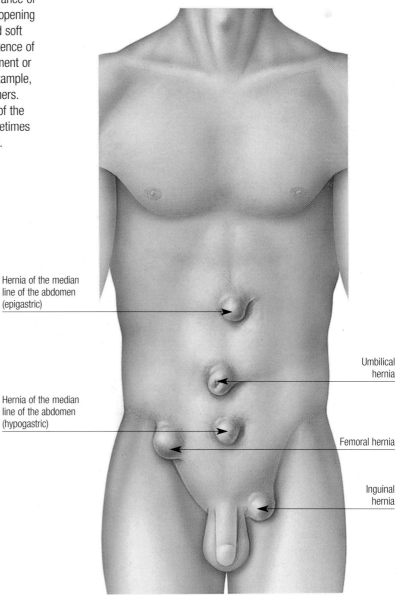

Hernia of the median line of the abdomen (epigastric)

Umbilical hernia

Hernia of the median line of the abdomen (hypogastric)

Femoral hernia

Inguinal hernia

### Peritoneum

The peritoneum is a rather extensive serous membrane basically made up of connective tissue that lines the interior of the walls of the abdominal cavity. It is bent back to cover the major portion of the organs contained inside it. Although the membrane is uninterrupted, it is made up of two layers or sheets: the parietal peritoneum and the visceral peritoneum. A space is created between the two layers of the peritoneum that contains only a thin film of a lubricating fluid composed of water, certain cells, and mineral substances. The fluid's primary function is to permit the movement of the sheets without causing friction between them and indirectly between the abdominal organs and the abdominal wall.

# Nutrients and nutritional needs

**Nutrients are basic substances contained in food that the body** regularly needs to form and conserve its tissues and to obtain energy. We must consume nutrients. The body uses them in physiological activities and to regulate metabolism correctly.

## Functions of nutrients

Each type of nutrient is used by the body in a particular way. Generally, nutrients have three types of functions:

■ **Structural function:** Nutrients are used to construct or repair tissues and organs. All of the proteins and some minerals have a structural function.

■ **Energy function:** Nutrients provide the energy necessary for the metabolic reactions required to live, for maintaining body heat, for muscle contractions and movement, and for many other purposes. Carbohydrates, fats, and proteins all have an energy function.

■ **Regulatory function:** Nutrients modulate the metabolic reactions and the activity of different organs. Various minerals and vitamins have a regulatory function.

## Chemical composition in the human body

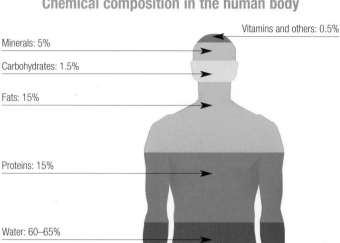

Vitamins and others: 0.5%

Minerals: 5%

Carbohydrates: 1.5%

Fats: 15%

Proteins: 15%

Water: 60–65%

It is believed that there are six classes of nutrients, each type having its own specific functions, but all of them are indispensable: carbohydrates, proteins, fats, vitamins, minerals, and water.

## Food groups

**Fruits**
contain water as well as important vitamins and a variable percentage of sugars and fats.

**Fats and sweets**
are characterized by their high energy potential. They are good diet complements, provided they are consumed in moderation.

**Milk and milk derivatives**
contain an important amount of various elementary nutrients. These are the most complete food items.

**Vegetables**
offer a minimum energy potential while, at the same time, providing a wide variety of minerals and vitamins that are indispensable regulatory elements so that organic metabolism may function correctly.

**Meats, fish, and eggs**
supply high-quality proteins, nutrients that the body requires to form and regenerate its tissues.

**Cereals, tubers, and legumes**
are noted for their especially high content of complex carbohydrates, the principal source of energy for the body.

*To be able to determine the qualities of the different foods and to lay the foundation for a healthy diet, it is best to classify foods into groups made up of products with an equivalent nutritional composition.*

## Calories released by the digestion of the different nutrients

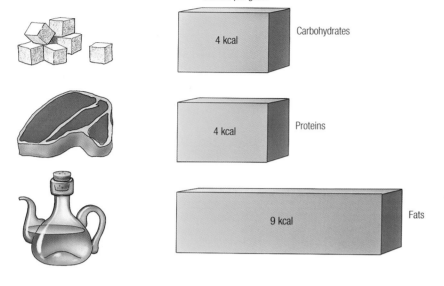

Calories per gram

4 kcal — Carbohydrates

4 kcal — Proteins

9 kcal — Fats

### Units

To calculate the energy of nutrients, we usually employ the unit of thermal or calorific energy called the calorie (with a capital letter, abbreviated Cal) and commonly referred to as kilocalorie or kcal. One kcal is the quantity of heat necessary to raise the temperature of 1 liter of distilled water from 14.5° C to 15° C. One calorie, abbreviated cal (with a small letter), consists of one-thousandth kcal. Additionally, another unit, the joule (J) and its multiple, the kilojoule (kJ), have been used.

**Unit equivalency**

| | |
|---|---|
| 1 cal: | 4.18 J |
| 1 kcal(Cal): | 1,000 calories |
| 1 kcal: | 4.18 kJ |
| 1 kJ: | 0.24 cal |

Many of the nutrients obtained from food are used by the body to obtain the energy necessary to develop the biochemical reactions of metabolism, to maintain body heat, and for muscular work. Only three types of basic nutrients are useful for this purpose: carbohydrates, proteins, and fats. Their combustion releases variable quantities of energy.

The body requires a basic quantity of energy for basal metabolism, that is, the fixed energy expenditure necessary to maintain permanent activity, to renew all of the tissues, and to maintain body temperature. In general, it is agreed that this expenditure amounts to 25 Cal per kilogram of weight during the day, although it varies as a function of body weight and size. It is slightly greater in men than in women. It is even greater in children—when growth is at its peak, during pregnancy—to meet the demands of the fetus, and during nursing—to make milk. In addition, caloric demands are affected by the level of physical activity. The more the muscles work, the more calories the body requires.

## Energy needs according to age, sex, weight, and physical activity

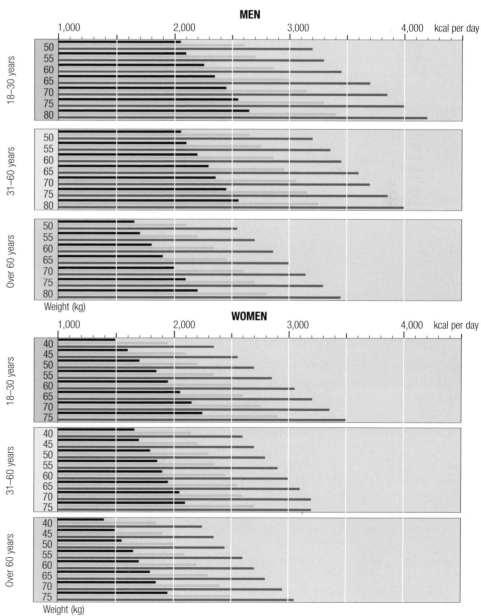

Light activity

Moderate activity

Heavy activity

# Carbohydrates

**Carbohydrates, also called sugars, are nutrients present in** almost all foods, primarily in fruits and grains. The body uses carbohydrates primarily to obtain the energy needed to perform the multiple biochemical reactions of metabolism. They are the principal fuel of the body.

## Structural formulas of monosaccharides

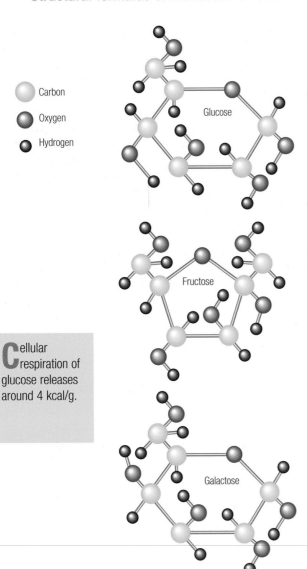

- Carbon
- Oxygen
- Hydrogen

Glucose

Fructose

Galactose

**C**ellular respiration of glucose releases around 4 kcal/g.

## Structural formulas of disaccharides

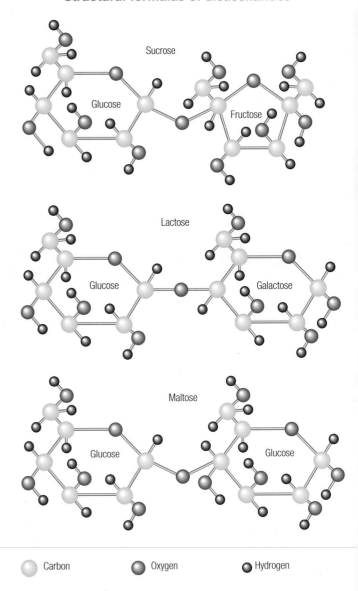

Sucrose

Glucose

Fructose

Lactose

Glucose

Galactose

Maltose

Glucose

Glucose

- Carbon
- Oxygen
- Hydrogen

## Types of carbohydrates

Carbohydrates are made up of atoms of carbon, oxygen, and hydrogen. Carbohydrates always have two hydrogen atoms for every oxygen atom, the same proportion found in water ($H_2O$). The basic unit of a carbohydrate is a monosaccharide. Carbohydrates are classified based on their chemical structure and the number of monosaccharide units from which they are formed. Simple carbohydrates, also called sugars, can be either monosaccharides or disaccharides. Glucose, fructose, and galactose are examples of monosaccharides. Disaccharides are formed from two monosaccharides. Examples of disaccharides include sucrose (table sugar), lactose (milk sugar), and maltose. Sucrose consists of one molecule of glucose and one of fructose. Lactose consists of one molecule of glucose and one of galactose. Maltose consists of two molecules of glucose. Complex carbohydrates, also called polysaccharides, are made up of numerous monosaccharides linked in long chains. Examples include the starches found in vegetables and the glycogen found in animals.

*The cereals (corn flakes, potatoes, and legumes) are foods rich in carbohydrates, the energy base of human nutrition.*

# Digestion of Carbohydrates

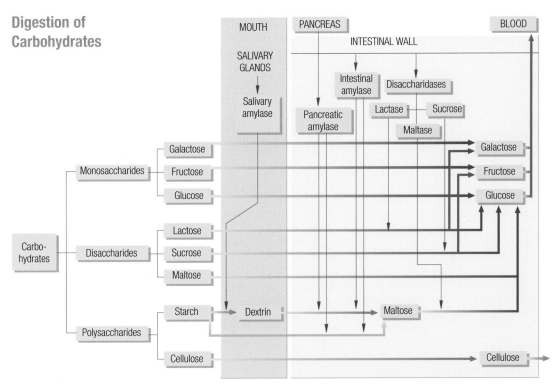

Most carbohydrates that are present in food are disaccharides and polysaccharides. However, only the tiny monosaccharides can be absorbed through the walls of the digestive system without difficulty. Therefore, to be absorbed, almost all carbohydrates must be digested by enzymes that break them up into their basic units, monosaccharides. At the end of the digestive process, absorption of the molecules of glucose, fructose, and galactose takes place in the intestines. Molecules are transported to the liver, where fructose and galactose are transformed into glucose, which is then released into the blood to be distributed throughout the body.

## Carbohydrate content

Per 100 grams of food

| | |
|---|---|
| Refined sugar | 100 g |
| Cornflakes | 84 g |
| Honey | 77g |
| White rice | 77 g |
| Wheat flour | 75 g |
| Food pasta | 73 g |
| Cookies | 73 g |
| Jellies | 70 g |
| Oats | 65 g |
| Pastries | 60 g |
| Beans | 60 g |
| White bread | 55 g |
| Whole-wheat bread | 49 g |
| Bananas | 21 g |
| Cooked potato | 20 g |
| Grapes | 17 g |

## Fiber content
(g/100 g of food)

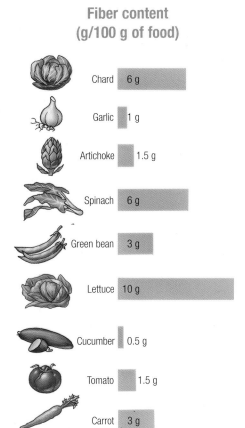

| | |
|---|---|
| Chard | 6 g |
| Garlic | 1 g |
| Artichoke | 1.5 g |
| Spinach | 6 g |
| Green bean | 3 g |
| Lettuce | 10 g |
| Cucumber | 0.5 g |
| Tomato | 1.5 g |
| Carrot | 3 g |

## Glucose

Cells of the human body can use only a single carbohydrate as an energy source—glucose. The molecules of glucose that are absorbed in the intestine through the breakdown of complex carbohydrates and that are released through the liver by conversion into monosaccharides circulate through the blood to reach the cells. They are then subjected to a chemical process of respiration that involves the release of energy. Glucose is so important that when blood tests are performed, its concentration is always measured. It is called glycemia or blood sugar since it is an indispensable indicator in determining a person's state of health.

## Food Nutrients

All foods contain carbohydrates except those that are made exclusively of fats, such as oils. Foods that are richest in carbohydrates are cereals and their derivatives, legumes, tubers, fruits, and, of course, common sugar in its different varieties, as well as honey and all sweet products.

## Cellulose: Vegetable Fiber

Cellulose is a complex carbohydrate that forms the walls of vegetable cells. Herbivorous animals have digestive enzymes that are capable of digesting this compound and thus releasing its components, glucose molecules, which are absorbed and used to exploit their energy potential. On the other hand, humans do not have the enzymes necessary to digest cellulose. It is a part of what we call "vegetable fiber" that is eliminated. Its consumption is beneficial, however, because it provides a larger volume of digestive residues—fiber. This means that the walls of the large intestine work better.

# Proteins

**Proteins are basic components of the body and must be** consumed daily. The body uses proteins to form and repair tissues and to regulate all metabolic processes.

## Tripeptide

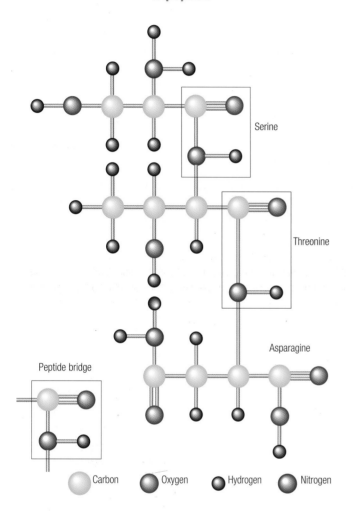

Serine

Threonine

Asparagine

Peptide bridge

Carbon  Oxygen  Hydrogen  Nitrogen

## Chemical structure of proteins

Proteins always contain carbon, oxygen, hydrogen, and nitrogen. Sometimes they also contain sulfur and phosphorus. The basic unit of a protein is an amino acid. Amino acids link together with peptide bonds to form chains. When only a few amino acids are bound together, the grouping is called a peptide. For instance, a dipeptide contains only two amino acids and a tripeptide contains only three. When the chain contains many amino acids, it is called a protein. Some proteins contain about 100 amino acids, and others are composed of more than 1,000.

Cellular respiration of proteins releases around 4 kcal/g.

## Amino acids

| Nonessential amino acids | Essential amino acids |
|---|---|
| Glutamic acid | Arginine |
| Alanine | Phenylalanine |
| Asparagine | Histidine |
| Cysteine | Isoleucine |
| Cystine | Leucine |
| Glycine | Lysine |
| Hydroxyproline | Methionine |
| Proline | Threonine |
| Serine | Tryptophan |
| Tyrosine | Valine |

## Types of amino acids

All proteins found in nature are from a combination of only 20 different amino acids, each one having its own chemical structure. The body must have all of the amino acids in order to form its own proteins. The body can synthesize some amino acids—the nonessential ones. However, it must obtain the essential amino acids by consuming them daily. Therefore, a person must eat a variety of foods that can provide all of the essential amino acids. Eating meat easily accomplishes this. However, consuming specific food combinations, such as beans and rice, provides all the essential amino acids to a vegetarian or vegan.

## Some amino acids

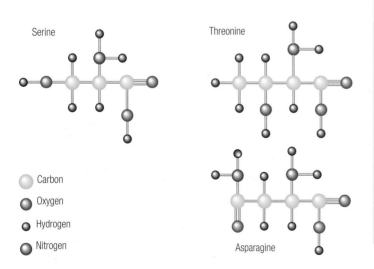

Serine

Threonine

Asparagine

Carbon

Oxygen

Hydrogen

Nitrogen

## Digestion of proteins

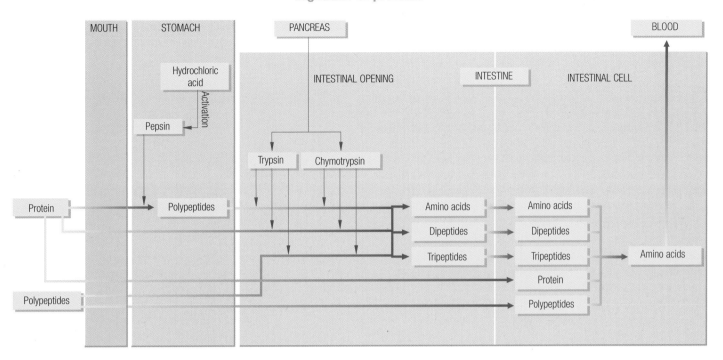

The digestion of proteins in foods begins in the stomach from the action of gastric juices. Hydrochloric acid that is secreted by the stomach mucosa activates an enzyme called pepsin that acts on proteins and breaks some of the bonds, releasing smaller polypeptide chains. When food passes into the small intestine, some enzymes processed by the pancreas release amino acids, dipeptides, and tripeptides that are absorbed by the cells of the intestinal walls. The breakdown occurs inside the walls in such a way that only free amino acids are released into the blood circulation. Subsequently, once distributed throughout the body, various amino acids will be combined with each other to form organic proteins.

## Protein content (approximately per 100 g of food)

| | | | |
|---|---|---|---|
| Parmesan cheese | 34 g | Chicken | 21 g |
| Soybean, grain | 34 g | Almonds | 20 g |
| Grains | 32 g | Beef (sirloin) | 20.5 g |
| Fresh tuna | 27 g | Lamb (leg) | 19 g |
| Peanuts | 24 g | Pig | 17 g |
| Lentils | 24 g | Sole | 16 g |
| Roquefort cheese | 23 g | Pork chop | 15 g |
| Broad beans | 23 g | Hazelnuts | 14 g |
| Duck | 22 g | Oats | 13 g |
| Rabbit | 22 g | Egg, whole | 13 g |
| Peas | 22 g | Wheat flour | 9.5 g |
| Sardines | 21 g | White bread | 7 g |
| Shrimp | 21 g | Whole cow's milk | 3.5 g |
| Cocoa powder | 21 g | | |

## Food sources

Practically all foods contain proteins with the sole exception of those that are formed only by fats, such as oils. Products that are richest in proteins are meats, fish, eggs, milk and some milk derivatives, legumes, all dry fruits, and cereals and their derivatives. Vegetables and some fruits also contain proteins although in a much smaller proportion.

## Functions of proteins

Proteins primarily have a structural function because they form part of the structure of the cell membranes and also because they constitute the framework that supports the tissues and organs of the body. They are also present in the intracellular fluid and in the cell nucleus. Proteins also make up enzymes, antibodies, some hormones, and an infinite number of elements that perform specific and highly varied actions, all of which are important in the correct functioning of the body. Finally, proteins can also be used to obtain energy.

# Fats

## Chemical structure of fats

Fats are composed of atoms of carbon, oxygen, and hydrogen combined in such a way that these nutrients will be insoluble in water. The most common lipids are triglycerides, made up of one molecule of the alcohol glycerol and three molecules of fatty acids. Since there are around 40 different fatty acids, the possible combinations are numerous. Fatty acids, made up of a long chain of carbon atoms each bonded to other carbon atoms and two atoms of hydrogen, can be divided into different types. If carbon atoms are linked to the maximum possible number of hydrogen atoms they are saturated because it is no longer possible for them to bond with any other hydrogen atom. If they have some free bonds, they are unsaturated. They are monounsaturated when only one bond is free and polyunsaturated when various bonds are free.

## Fatty acids

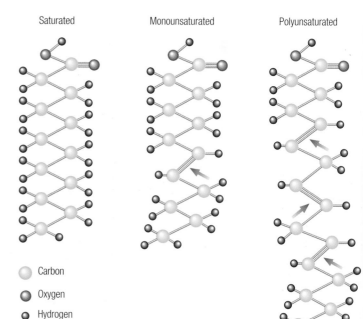

Saturated    Monounsaturated    Polyunsaturated

○ Carbon
◉ Oxygen
● Hydrogen
→ Double bond

## Triglycerides

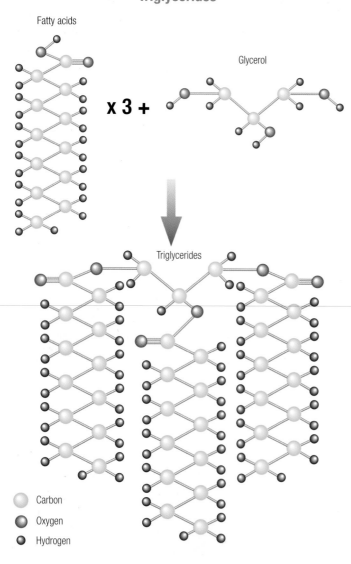

Fatty acids

Glycerol

x 3 +

Triglycerides

○ Carbon
◉ Oxygen
● Hydrogen

Cellular respiration of fats releases around 9 cal/g.

## Cholesterol

Cholesterol is not an essential nutrient because the body can synthesize it in the liver. Its functions are widely varied and important. For example, it is part of the cellular membranes and is the precursor of various hormones. However, consuming excessive amounts of cholesterol is dangerous. High cholesterol levels in the blood lead to a major risk of cardiovascular diseases, such as atherosclerosis and myocardial infarction—the main causes of mortality in industrialized countries. Foods that are richest in cholesterol are egg yolks, butter, and animal fats.

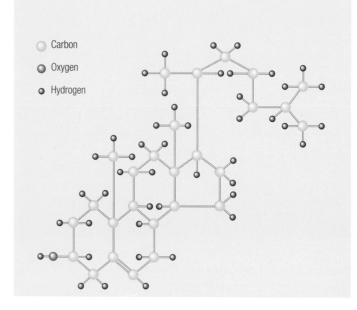

○ Carbon
◉ Oxygen
● Hydrogen

## The digestion and absorption of fats

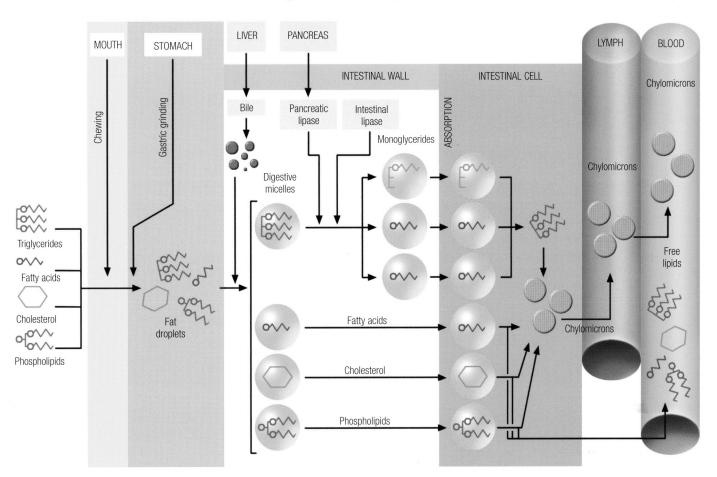

The fats contained in food reach the intestine in the form of small drops that cannot be attacked by digestive enzymes. In the duodenum, bile produced in the liver acts on these drops and exerts an emulsifying effect, breaking them down into microscopic particles called digestive micelles. The pancreatic and intestinal enzymes in the lipase family act upon them. These enzymes release gastric acids to penetrate the intestinal cells. There the micelles are regrouped to form particles known as chylomicrons, which are soluble in organic liquids. The chylomicrons then move on to the lymph vessels of the intestinal villi and by lymphatic circulation reach the blood circulation to be distributed throughout the body.

| **Fat content** | |
|---|---|
| (approximately per 100 g of food) | |
| Oils | 100 g |
| Butter | 83 g |
| Margarine | 83 g |
| Mayonnaise | 78 g |
| Bacon | 70 g |
| Coconut | 60 g |
| Hazelnuts | 60 g |
| Peanuts | 60 g |
| Nuts | 60 g |
| Almonds | 54 g |
| Fried potatoes | 37 g |
| Chocolate milk | 34 g |
| Swiss cheese | 33 g |
| Cream | 30 g |
| Avocado | 16 g |
| Pork | 25 g |
| Beef | 20 g |

## Content of saturated and unsaturated fats

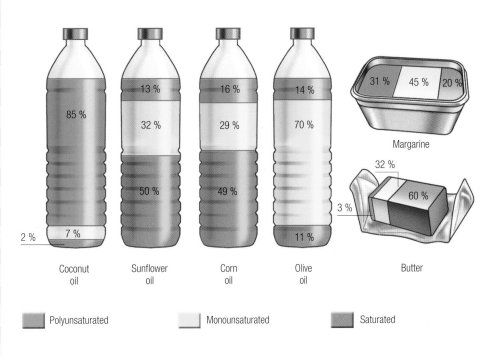

# Vitamins

**Vitamins are widely diverse chemical substances that the body** cannot synthesize. Therefore, it needs to incorporate them in small quantities from food to ensure their correct functioning. Vitamins perform regulatory functions in multiple essential metabolic processes.

## Fat-soluble vitamins

| Name | Functions | Daily needs | Sources |
|---|---|---|---|
| **Vitamin A or retinol** | Participates in sight and in the growth, reproduction, and maintenance of the epithelial tissue (skin and mucosae) | children: 0.4–0.6 mg; men: 4 mg; pregnancy: 1 mg; nursing 1.2 mg | Milk and derivatives (butter), liver, egg yolk, oily fish, vegetables rich in carotenes, such as carrots, squash, and green-leaf vegetables |
| **Vitamin D or calciferol** | Participates in the regulation of metabolism of calcium and phosphorus and in muscle activity; it is indispensable in the growth process | children: 10 mg; men: 7.5 mg; pregnancy: 10 mg; nursing: 10 mg | Liver, fish and shellfish, meat, milk and milk derivatives, eggs; it is produced by the skin due to the influence of the sun's rays |
| **Vitamin E or tocopherol** | Antioxidant action participates in maintenance of cell membranes | children: 7.5–10 mg; men: 5–8 mg; pregnancy: 12–15 mg; nursing: 12–15 mg | Eggs, seed oils |
| **Vitamin K or menaquinone** | Required by the liver to help process coagulation factors needed to stop hemorrhages | approximately 1 mg, but consuming it is not necessary because intestinal bacterial flora synthesize it | Liver, kidneys, vegetables and fruits; it is synthesized by the bacteria of the intestinal flora |

Although each vitamin has its own specific name, the different vitamins usually are labeled with letters of the alphabet and subscripts. Vitamins were named gradually as they were discovered, even when their chemical formulas were not yet known.

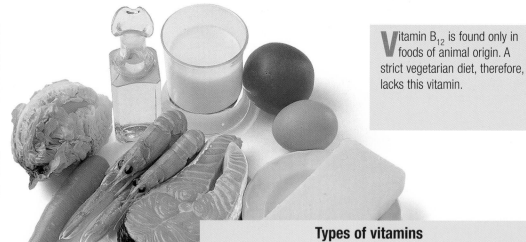

*Some foods that are sources of fat-soluble vitamins.*

Vitamin $B_{12}$ is found only in foods of animal origin. A strict vegetarian diet, therefore, lacks this vitamin.

## Types of vitamins

Vitamins are classified in two major groups, depending on their solubility properties.

■ **Water-soluble vitamins** include the B-complex vitamins and vitamin C. If these vitamins are consumed in excess, the surplus is eliminated by the kidneys and voided in the urine. Excess consumption of water-soluble vitamins does not cause any health problems.

■ **Fat-soluble vitamins** include vitamins A, D, E, and K. These vitamins are found only in foods that contain a certain quantity of lipids. Additionally, the intestines require the presence of fats to absorb them. Since they tend to be deposited in the fatty tissues, consuming fat-soluble vitamins in excess will cause them to accumulate in the body. The disorder hypervitaminosis can then result. The specific manifestations of the disorder depend on the specific vitamin involved.

## Conservation

Vitamins present in foods will deteriorate partially as a result of prolonged exposure to light, contact with the air, and cooking. Therefore, to ensure their vitamin value, one must protect these foods against light and keep them in a cool place. One should preferably consume raw vegetables and fruit.

## Water-soluble vitamins

| Name | Functions | Daily needs | Sources |
|------|-----------|-------------|---------|
| **Vitamin B$_1$ or thiamine** | Participates in carbohydrate metabolism and the activity of the peripheral nerves, the heart, and the intestines | children: 0.3–1 mg; adults: 1.3–1.5 mg | Whole cereals or derivatives, yeast, milk, egg, beef, pork, dried fruits, legumes, and vegetables in general |
| **Vitamin B$_2$ or riboflavin** | Participates in processes of respiration and intracellular oxidation; acts as coenzyme in the metabolism of carbohydrates, proteins, and fats; involved in hemoglobin metabolism | children: 0.6–2.5 mg; adults: 1.6–1.8 mg; nursing: 2 mg | Liver, kidney, yeasts, milk and milk derivatives, egg, dried fruit, whole cereals, vegetables, and fruits |
| **Niacin or nicotinamide** | Participates in the metabolism of carbohydrates, proteins, and fats | children: 8–11 mg; men: 18 mg; women: 13 mg; pregnancy: 5 mg; nursing: 15 mg | Red meats, fish, poultry, legumes, whole cereals and their derivatives |
| **Vitamin B$_5$ or pantothenic acid** | Constituent of coenzyme A; participates in metabolism of carbohydrates, proteins, and fats as well as in the synthesis of numerous substances | 5–10 mg | Present in practically all foods |
| **Vitamin B$_6$ or pyroxidine** | Participates in protein metabolism and in the formation of blood | children: 0.4–1.4 mg; adults: 1.8 mg; pregnancy: 2.5 mg | Yeast, red meats, fish, poultry, soy, whole cereals and their derivatives, dried fruit, some fruits |
| **Vitamin B$_8$, biotin or vitamin H** | Participates in the metabolism of carbohydrates, proteins, and fats; stimulates the growth of nervous tissue and the skin | 10 μg | Yeast, kidney, liver, egg, chocolate, mushrooms |
| **Vitamin B$_9$ or folic acid** | Participates in the maturation of the red blood cells and in the cell division process; indispensable for the formation of new tissues during the growth phase | 200–300 μg | Liver, legumes, cereals, soy, milk, meats, dried fruit, green-leaf vegetables, all fresh fruits |
| **Vitamin B$_{12}$ or cobalamine** | Participates in the metabolism of carbohydrates, proteins, and lipids involved in the maturation of red blood cells, in the synthesis of DNA, and in the activity of the nervous system | 2.5–4 mg | Liver, kidney, meats, fish, meat and milk derivatives, egg (absent in vegetable foods) |
| **Vitamin C or ascorbic acid** | Participates in intracellular metabolism; indispensable for the synthesis of collagen; protects the skin and mucosa; participates in the formation of some hormones and facilitates the intestinal absorption of iron | children: 35–40 mg; adults: 60 mg; pregnancy: 80–100 mg | Fruits (citrus, kiwi, pineapple, strawberries) and fresh vegetables (pimento, broccoli, cabbage, watercress, chard, potato) |

*Some foods are sources of water-soluble vitamins.*

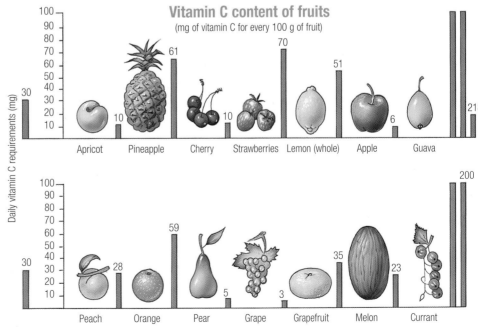

**Vitamin C content of fruits**

(mg of vitamin C for every 100 g of fruit)

Apricot 10 · Pineapple 61 · Cherry 10 · Strawberries — · Lemon (whole) 70 · Apple 51 · Guava 6 · 218

Peach 28 · Orange — · Pear 59 · Grape 5 · Grapefruit 3 · Melon 35 · Currant 23 · 200

Daily vitamin C requirements (mg): 30

# Minerals

**Minerals are inorganic chemical substances that are present in** varying proportions in food. They are considered nutrients because the body must incorporate them in varying quantities, both to form its structures and because they contain elements that regulate multiple metabolic processes.

## Functions, requirements, and sources of principal minerals

| Minerals | Principal functions | Daily requirements | Sources |
|---|---|---|---|
| **Calcium** | Part of the structure of bones and teeth; calcium participates in nerve impulses, muscle contraction, and blood coagulation | 0–12 months: 500–600 mg; 1–10 years: 600–800 mg; 11–18 years: 1,000–1,200 mg; adults: 800–1,000 mg; pregnant and nursing women: 1,200 mg | Milk and milk derivatives, dried fruit, legumes, egg, fish, shellfish, vegetables |
| **Sodium** | Participates in the regulation of body fluids and arterial blood pressure as well as in the transmission of nerve impulses and muscle contractions | From 1 to 3 g, depending on factors such as body activity, kidney function, or perspiration | Table salt, sausages, cheeses, canned food, meats, fish, vegetables, and fruits |
| **Potassium** | Acts with sodium in the transmission of nerve impulses and in the maintenance of the water-salt balance; it is essential in the metabolism of sugars and proteins | 0–6 months: 500 mg; 6–12 months: 700 mg; 1–2 years: 1,000 mg; 2–5 years: 1,400 mg; 6–10 years: 1,600 mg; adults: 2,000 mg | Fruits and vegetables, meats, fish and shellfish, legumes, cereals, dried fruit |
| **Phosphorus** | Part of the structure of bones and teeth, it is a constituent of the cell membrane and the chromosomes, participating in the procurement of energy and in muscle contraction | 0–12 months: 200–400 mg; 1–10 years: 600–800 mg; 11–18 years: 800–1,000 mg; adults: 600–800 mg; pregnant and nursing women: 800–1,000 mg | Milk and milk products, dried fruit, cereals, legumes, eggs, meats, fish, and shellfish |
| **Magnesium** | Part of the bones, magnesium activates intracellular enzymes, participating in the transmission of impulses to the muscle tissue | 0–12 months: 40 mg; 1–10 years: 60–170 mg; adolescents: 300–400 mg; adults: 350 mg | Milk, green-leaf vegetables, meats, soy, cocoa, dried fruits, marine products. |
| **Iron** | Part of the hemoglobin of the red blood cells, of the myoglobin of the muscles, and of multiple enzymes involved in metabolism | children: 10–15 mg; adult males: 10–12 mg; adult women: 15–18 mg; pregnancy: 15–20 mg. | Liver, red meats, poultry, eggs, legumes, dried fruits, some vegetables |
| **Fluorine** | Part of bones and teeth that protects against decay | 1–4 mg | Fluorinated water, fluorinated table salt, marine products, some varieties of tea |
| **Iodine** | Part of the hormones processed by the thyroid glands, which regulate the general metabolism of the organism, iodine has a function in the process of growth and maturation of the nervous system | children: 100 µg; adults: 125–150 µg | Marine products: fish, mollusks, crustaceans, algae |
| **Copper** | Part of numerous enzymes, copper participates in the synthesis of proteins and in the procurement of energy | children: 0.5–2 mg; adults: 2–5 mg | Fish, crustaceans and mollusks, soy, legumes, dried fruits, whole cereals |
| **Zinc** | Important component of various enzymes, indispensable for growth, participates in synthesis of proteins and nucleic acids | 0–12 months: 3–5 mg; 1–10 years: 10 mg; adults: 10–15 mg | Fish and crustaceans, milk and its derivatives, eggs, meats, legumes, whole cereals |
| **Manganese** | Participates in multiple enzyme reactions and in the synthesis of lipids and mucopolysaccharides | 2–5 mg | Marine algae, milk products, legumes, cereals, meats. |
| **Sulfur** | Component of various amino acids, participates in cellular respiration mechanisms and in the procurement of energy | 5 µg | Cereals, fish and marine products, milk products, meats, legumes, egg, dried fruits |

## Types of minerals

The body consists of widely diverse minerals that are part of its structure or that are found in compounds that perform various activities. Minerals account for 5 to 6 percent of body weight. They are essential for growth. Since a portion is eliminated along with waste products and secretions, it is necessary throughout an entire lifetime to restore them in a proportion equivalent to the losses. In the field of nutrition, minerals are classified into two major groups, depending on their daily requirements. Some are included in the group of macronutrients because their content in the body is important. Therefore, a considerable regular supply is required through food. This group includes calcium, iron, phosphorus, sodium, potassium, and magnesium. Others are included in the group of micronutrients, also called oligoelements, because their body content is very small. Only a small amount must be supplied by the food regularly to meet the needs. This group includes selenium, fluorine, iodine, manganese, copper, molybdenum, zinc, chromium, cobalt, nickel, and vanadium, among others.

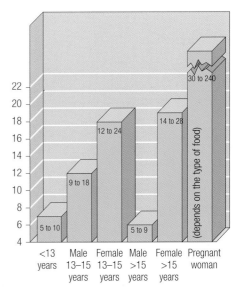

**Daily iron requirements** (mg/day)

*Vegetables are foods rich in minerals, and therefore, they must be a part of the daily diet.*

**Daily calcium requirements**
(mg/day)

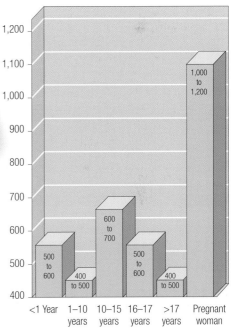

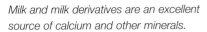

*Milk and milk derivatives are an excellent source of calcium and other minerals.*

## Mineral Supplements

All minerals required by the body can be obtained in sufficient quantities by eating a varied and balanced diet. Minerals supplements are needed only during specific times in life when the body's mineral needs are higher than usual or when the body has a mineral deficiency. A physician should prescribe a specific mineral supplement. In general, a person should not take a multivitamin/mineral supplement without first consulting a physician. An excess intake of minerals can be more harmful than a mineral deficiency.

# Water

**Water is a fundamental element for life. It is the principal** component of the human body and of all living things. It is the basic element of organic matter, the medium where all biochemical reactions of the body take place. The body can survive only a few days without the proper intake of water.

## Water molecule

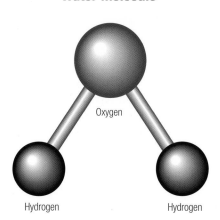

Oxygen

Hydrogen    Hydrogen

## Proportion of water (% of body weight)

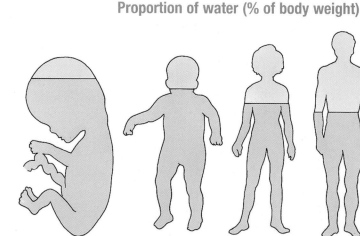

Fetus 90%    Newborn 80%    Child 70%    Adult 60–64%    Senior citizen 55%

*Pure water is a colorless, odorless, and tasteless liquid made up of two chemical elements: oxygen (O) and hydrogen (H). A molecule of water is made up of one atom of oxygen and two atoms of hydrogen, which is why its chemical formula is $H_2O$.*

*Water is important in the body in terms of quality of life since it is the principal component of the body. More than half of the body mass of a human being is made up of water, although the proportion, much higher at the beginning of life (around 80% in the newborn), declines gradually throughout an individual's lifetime.*

An individual can survive only about a week, or at most ten days, without consuming any water. Recorded cases where this limit has been exceeded are exceptional.

## Fluid contained within the body

Water is found both inside the cells and in the intercellular spaces and certain specific compartments, such as the cardiovascular system. Two-thirds of the total, in other words, 66%, of the body's water is contained inside the cells, called the intracellular fluid. Another 25% of the total is distributed between the cells of the tissues. This is the intercellular fluid. The remainder, less than 10%, consists of extracellular fluid that is part of the blood, lymph, and other organic secretions.

Water

Intracellular water 66%

Intercellular water 25%

Water in blood, lymph, and other organic fluids 9%

## Water balance of the body

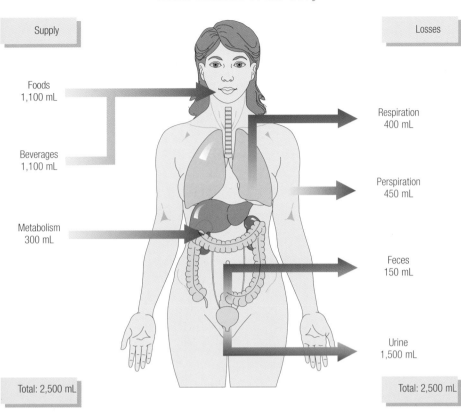

| Supply | | Losses | |
|---|---|---|---|
| Foods 1,100 mL | | Respiration 400 mL | |
| Beverages 1,100 mL | | Perspiration 450 mL | |
| Metabolism 300 mL | | Feces 150 mL | |
| | | Urine 1,500 mL | |
| Total: 2,500 mL | | Total: 2,500 mL | |

The body loses water constantly, which is why it is necessary to take in an equivalent quantity to maintain stable levels and to ensure water balance. Losses occur in various ways: through waste in urine and feces, through the skin, through perspiration, and through the lungs as a result of respiration. The chemical reactions that take place in the body to metabolize carbohydrates, proteins, and fats creates a certain gain in water, so-called endogenous water, around 300 mL per day in adults. This volume is not enough to make up for the losses. It is therefore necessary to make up the difference by consuming water. Exogenous water is supplied through beverages, which basically consist of water, as well as through foods that contain water in varying degrees.

## Thirst

The human body has an alarm mechanism that alerts it to the fact that more water must be consumed. It is called thirst. If there is little water in the body, then there is a proportional decline in the quantity that forms the blood and, consequently, the substances that are dissolved in it will be more concentrated. Specifically, at the base of the cerebrum is a center that is connected to receptors that specialize in analyzing the concentration of substances in blood. If the center detects that the blood concentration is higher than normal, it generates stimuli that the cerebrum interprets as the particular sensation of thirst, a sensation that provides the impetus for drinking.

## Exceptional lubricant

The eyes, the joints, and the mucosae require water to reduce the friction that they endure every day. Thanks to water, for example, our eyes and our tongue remain constantly moist.

### Principal external sources of water

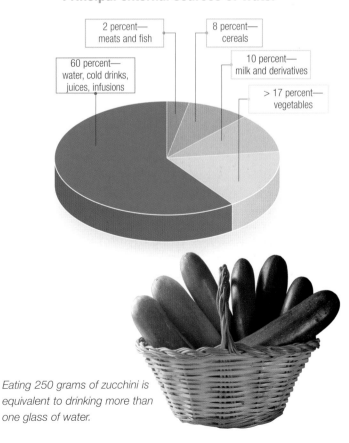

2 percent— meats and fish
8 percent— cereals
60 percent— water, cold drinks, juices, infusions
10 percent— milk and derivatives
> 17 percent— vegetables

*Eating 250 grams of zucchini is equivalent to drinking more than one glass of water.*

### Water content in some foods

| Pepper | 96% | Meat | 50–70% |
|---|---|---|---|
| Milk | 90% | Cheeses | 39–50% |
| Melon | 89.9% | Cookies | 5% |
| Fish | 73–84% | Nuts | 4% |

# Healthy nourishment

**Nourishment is considered healthy when, regardless of cultural** habits and individual preferences, it supplies the body with all the nutritional elements it requires in adequate quantities and proportions to meet an individual's basic needs without either shortages or excesses: a complete, balanced, and varied diet.

## Varied nourishment

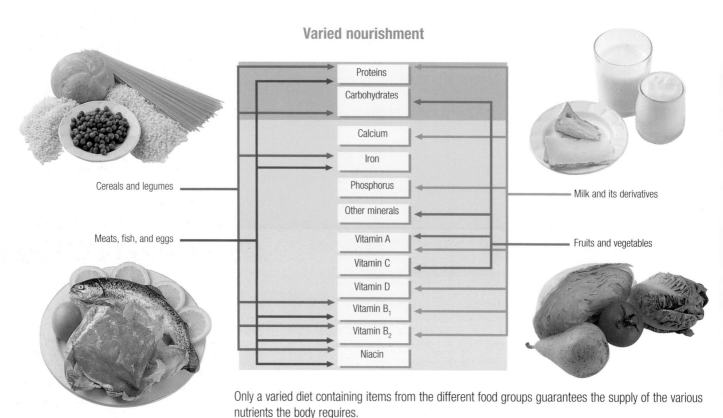

Cereals and legumes

Meats, fish, and eggs

Proteins
Carbohydrates
Calcium
Iron
Phosphorus
Other minerals
Vitamin A
Vitamin C
Vitamin D
Vitamin B$_1$
Vitamin B$_2$
Niacin

Milk and its derivatives

Fruits and vegetables

Only a varied diet containing items from the different food groups guarantees the supply of the various nutrients the body requires.

One essential requirement of healthy nourishment is that the meal must be complete. In other words, it must supply all types of nutrients, those used for structural purposes (protein), those used for energy purposes (carbohydrates and fats), and those used for regulatory functions (minerals and vitamins). Only consuming a variety of products from the different food groups will guarantee the supply of diverse nutrients that the body needs.

## Qualitative balance

To prevent shortages or excesses, the usual diet must be based on suitable proportions of different nutrients. To determine the proper proportions needed, the total calorie requirements of each person are calculated on the basis of age, size, weight, physical makeup, and activity. The requirements of each type of nutrient are met by complying with the percentages of carbohydrates, proteins, and fats indicated in the illustration. The supply of minerals and vitamins, in turn, can always be guaranteed when one eats a varied diet since the absolute quantities required are minimal and can be met through the consumption of diverse products.

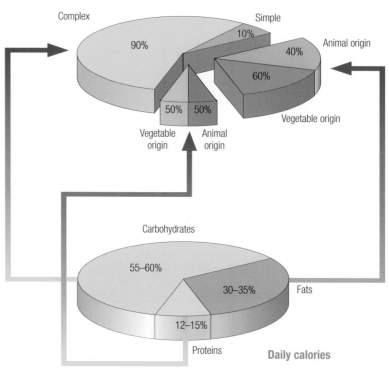

Complex

Simple

90%

10%

Animal origin

40%

60%

Vegetable origin

50%  50%

Vegetable origin   Animal origin

Carbohydrates

55–60%

30–35%   Fats

12–15%

Proteins

**Daily calories**

## Example of a quantitative balance

**Daily intake of energy**

Breakfast — 500 kcal

+

Dinner — 700 kcal

+

Supper — 900 kcal

+

Snacks during the day — 500 kcal

Daily total = 2,600 kcal

**Daily expenditure of energy**

Basal metabolism — 1,500 kcal

+

Sedentary activity — 400 kcal

+

Movements — 400 kcal

+

Leisure — 300 kcal

Daily total = 2,600 kcal

*A correct balance between energy supply and calorie expenditure makes it possible to keep the body functioning adequately without shortages or surpluses. This illustration shows an example of the correspondence between the consumption and expenditure of calories in an adult weighing approximately 130 pounds (60 kg).*

## Food rations

Cereals, tubers, and legumes

Milk and milk derivatives

Meat, fish, and eggs

Vegetables

Fruit

Nutritionists advise basing the calculations on the idea of the food ration: the usual quantity of a given food item that is usually consumed in a prepared dish, that is to say, the average ration of a certain meal that is usually served in a home or that one expects to get in a restaurant. This illustration shows examples of rations of different types of food: bread (one bread roll of 1.75 oz./50 g), rice (1 oz./30 g, raw), potatoes (3 oz./80 g), lentils or beans (3–3.5 oz./80–100 g, raw), milk (8 fl. oz./200 mL), yogurt (8.5 oz./240 g), dried cheese (1 oz./30 g), beef (steak, 3.5 oz./100 g), chicken (one quarter, 9 oz./250 g with bone), fish (4 oz./120 g), two eggs, ham (3 oz./80 g), salad (7 oz./200 g), green beans (3.5 oz./100 g), carrots (3.5 oz./100 g); apples (4.5 oz./130 g), orange (5.5 oz./160 g), bananas (3 oz./80 g), peaches (5 oz./140 g).

## Daily recommended food rations based on body weight

| Weight (lb.) | Cereals, legumes, and tubers | Milk and derivatives | Meats, fish, eggs | Greens and vegetables | Fruits |
|---|---|---|---|---|---|
| 110 | ▭▭▭▭ | ▭▭ | ▭▭ | ▭▭ | ▭▭ |
| 120 | ▭▭▭▭▭ | ▭▭ | ▭▭ | ▭▭ | ▭▭ |
| 130 | ▭▭▭▭▭▭ | ▭▭▭ | ▭▭ | ▭▭ | ▭▭ |
| 140 | ▭▭▭▭▭▭ | ▭▭▭ | ▭▭ | ▭▭ | ▭▭ |
| 155 | ▭▭▭▭▭▭ | ▭▭▭ | ▭▭ | ▭▭ | ▭▭ |
| 165 | ▭▭▭▭▭▭▭ | ▭▭▭ | ▭▭ | ▭▭ | ▭▭ |
| 175 | ▭▭▭▭▭▭▭▭ | ▭▭▭ ▪ | ▭▭ ▪ | ▭▭ ▪ | ▭▭ ▪ |
| 185 | ▭▭▭▭▭▭▭▭▭ | ▭▭▭ ▪ | ▭▭ ▪ | ▭▭ ▪ | ▭▭ ▪ |

▭ Food ration    ▪ Average food ration

# Eye

**The eye, also called the eyeball, is a complex and delicate** structure that constitutes a sensory organ. Its job is to catch the light stimulus coming from the outside in order to convert it into nerve impulses that are transmitted to the cerebrum, where they are turned into visual images that represent the elements that surround us.

## Eyeball cross section

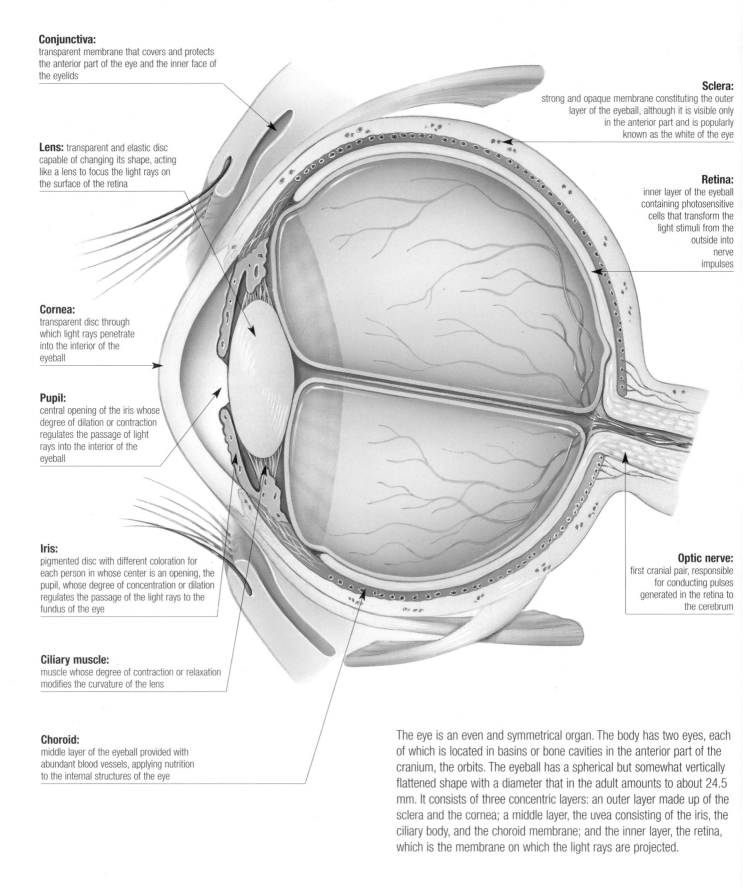

**Conjunctiva:**
transparent membrane that covers and protects the anterior part of the eye and the inner face of the eyelids

**Lens:** transparent and elastic disc capable of changing its shape, acting like a lens to focus the light rays on the surface of the retina

**Cornea:**
transparent disc through which light rays penetrate into the interior of the eyeball

**Pupil:**
central opening of the iris whose degree of dilation or contraction regulates the passage of light rays into the interior of the eyeball

**Iris:**
pigmented disc with different coloration for each person in whose center is an opening, the pupil, whose degree of concentration or dilation regulates the passage of the light rays to the fundus of the eye

**Ciliary muscle:**
muscle whose degree of contraction or relaxation modifies the curvature of the lens

**Choroid:**
middle layer of the eyeball provided with abundant blood vessels, applying nutrition to the internal structures of the eye

**Sclera:**
strong and opaque membrane constituting the outer layer of the eyeball, although it is visible only in the anterior part and is popularly known as the white of the eye

**Retina:**
inner layer of the eyeball containing photosensitive cells that transform the light stimuli from the outside into nerve impulses

**Optic nerve:**
first cranial pair, responsible for conducting pulses generated in the retina to the cerebrum

The eye is an even and symmetrical organ. The body has two eyes, each of which is located in basins or bone cavities in the anterior part of the cranium, the orbits. The eyeball has a spherical but somewhat vertically flattened shape with a diameter that in the adult amounts to about 24.5 mm. It consists of three concentric layers: an outer layer made up of the sclera and the cornea; a middle layer, the uvea consisting of the iris, the ciliary body, and the choroid membrane; and the inner layer, the retina, which is the membrane on which the light rays are projected.

## Conjunctiva

### Projection of conjunctiva over the eyelids

Conjunctiva

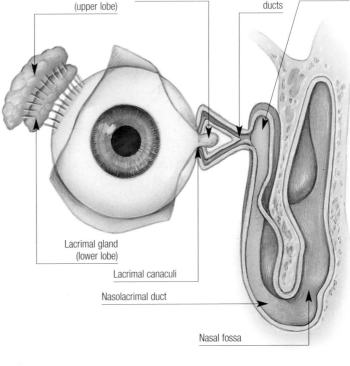

Conjunctiva

Upper eyelid

Sclera

Cornea

Lens

Iris

Lower eyelid

The conjunctiva is a thin, transparent, and mucous membrane that covers the anterior part of the eye, although it does leave the cornea free. It is folded back to line the eyelids on the inside. Its primary function is to provide protection against external objects.

## Lacrimal system

Lacrimal gland (upper lobe)

Lacrimal sac

Tear ducts

Tear sac

Lacrimal gland (lower lobe)

Lacrimal canaculi

Nasolacrimal duct

Nasal fossa

The lacrimal system consists of a gland situated in the upper and outer part of the ear that secretes a fluid intended to lubricate, nourish, and protect the anterior surface of the eyeball so the cornea will not dry out. This fluid is produced continuously and is distributed over the surface of the eye by the eyelid. Any excess is drained through excretory ducts that start in the inner corner of the eye and that empty into the corresponding nasal fossa.

## Eyelid (cross section)

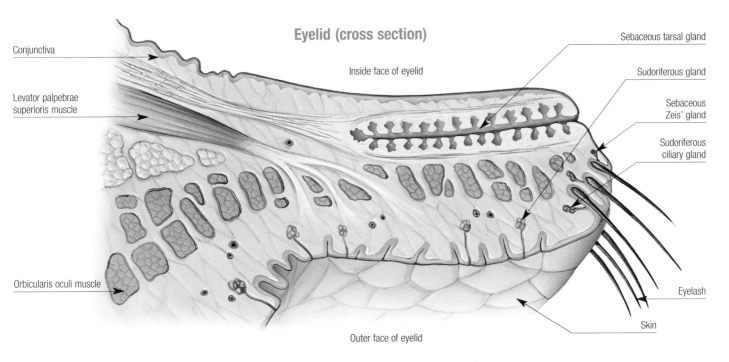

Conjunctiva

Levator palpebrae superioris muscle

Inside face of eyelid

Sebaceous tarsal gland

Sudoriferous gland

Sebaceous Zeis' gland

Sudoriferous ciliary gland

Orbicularis oculi muscle

Eyelash

Skin

Outer face of eyelid

The upper and lower eyelids are a kind of veil that, when closed, cover the eyeball completely. When open, they form a crack that leaves the cornea and a part of the sclera free. Covered with skin on the outside and lined on the inside by the conjunctiva, the eyelid harbors muscles within its thick layers, the orbicularis oculi and the levator palpebrae superioris, which are innervated by specific cranial nerves and responsible for the eyelid. There are also some sweat (sudoriferous) and sebaceous glands that empty onto the inner face and onto the edge where the eyelashes emerge.

# Vision

**Vision is the sense that provides most of our information.** It depends on complex mechanisms. The light stimuli that penetrate the eye and are transformed into nerve pulses must follow a path along the visual tracts up to the cerebrum, where they are processed and make us aware of images.

## Projection of images on the retina

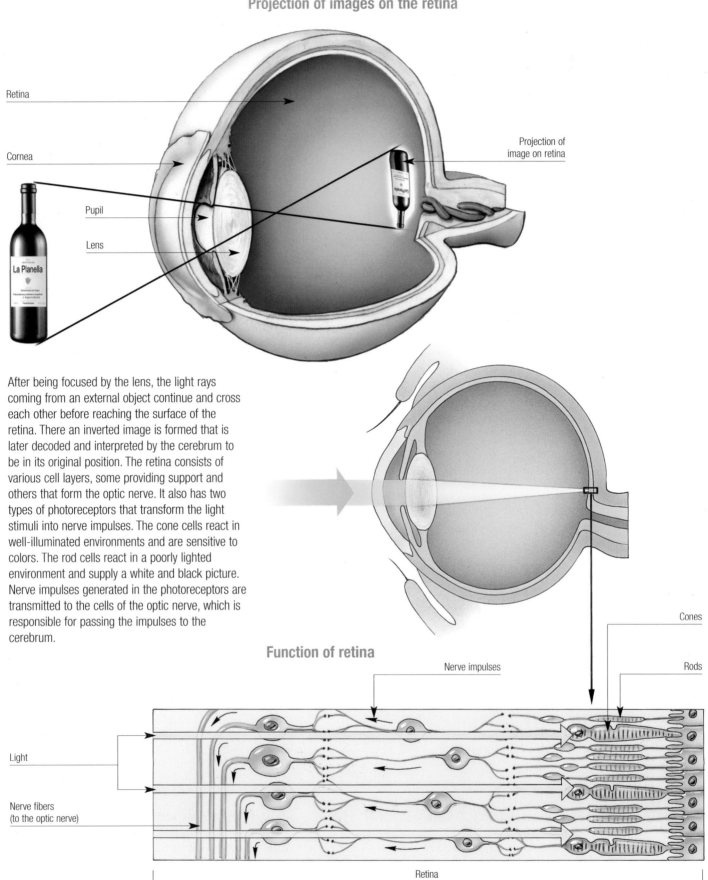

Retina

Cornea

Pupil

Lens

Projection of image on retina

After being focused by the lens, the light rays coming from an external object continue and cross each other before reaching the surface of the retina. There an inverted image is formed that is later decoded and interpreted by the cerebrum to be in its original position. The retina consists of various cell layers, some providing support and others that form the optic nerve. It also has two types of photoreceptors that transform the light stimuli into nerve impulses. The cone cells react in well-illuminated environments and are sensitive to colors. The rod cells react in a poorly lighted environment and supply a white and black picture. Nerve impulses generated in the photoreceptors are transmitted to the cells of the optic nerve, which is responsible for passing the impulses to the cerebrum.

## Function of retina

Cones

Rods

Nerve impulses

Light

Nerve fibers (to the optic nerve)

Retina

## Field of vision

Peripheral field of vision

Central field of vision

Binocular field of vision

Monocular field of vision (left eye)

Monocular field of vision (right eye)

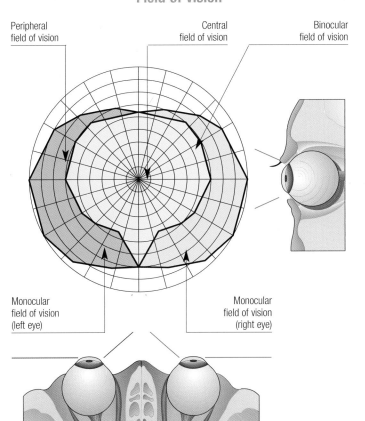

Each eye can only receive light rays coming from objects situated within a margin of space. Perception is limited in the upper part by the eyebrow, in the internal part by the nose, and in the external and lower parts by the borders of the eyeball. This shapes into a field of vision that in the horizontal direction covers 180° but in the vertical direction comprises 140°. In this space, we distinguish a central field of vision. It corresponds to the image projected on the macula lutea, the retinal zone with major concentration of photoreceptors, where vision is the sharpest. The peripheral field of vision covers the areas that are less populated with photoreceptors and where vision is not as sharp at a greater distance from the macula lutea.

When seeing with both eyes, the vision fields of both eyes are partially superimposed in the central part, the binocular vision field. There is a zone, however, that is perceived by only one eye and that remains beyond the reach of the other. It therefore corresponds to the monocular vision field. This is the reason the vision field varies when both eyes are open or when one eye is closed.

## Vision paths

Nerve impulses are produced in the photoreceptors of the retina as a result of the impact of the light rays on the retina. The nerve impulses travel along the optic nerve until they reach the brain, where the perceptions are noted. The more superficial cells of the retina are grouped and emerge through the posterior pole of the eyeball to form the optic nerve. The two optic nerves, each from one eye, cut across the lower face of the cerebrum and flow together at its base at a point close to the hypophysis. From there, where a part of the nerve fibers of both optic nerves cross each other, spring the optic tracts. These extend all the way to the optical thalamus, specifically the external geniculate bodies from which springs the optic radiation. These tracts reach all the way to the cerebral cortex of the occipital lobe, where the vision area is located. It is in this zone that the nerve impulses coming from the eyes are converted into visual sensations, thus creating conscious perceptions by mechanisms that are as yet little known.

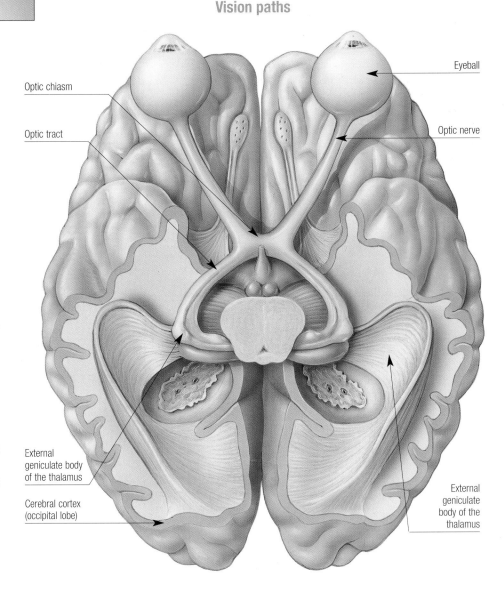

Optic chiasm

Optic tract

Eyeball

Optic nerve

External geniculate body of the thalamus

Cerebral cortex (occipital lobe)

External geniculate body of the thalamus

# Ocular refraction and its defects

**The optic system of the eye consists of various elements that** refract rays as they pass through different media. This refraction diverts the path of the rays to ensure that they form a clear image on the retina. Very common vision problems arise when these mechanisms fail.

## Lens accommodation mechanism

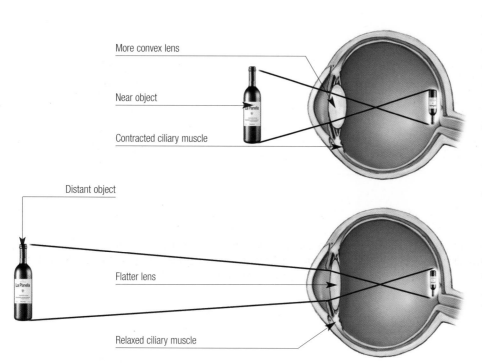

More convex lens

Near object

Contracted ciliary muscle

Distant object

Flatter lens

Relaxed ciliary muscle

To view objects correctly, their image must be formed exactly upon the retina. Otherwise, they will be perceived as blurry. The components of the optic system of the eye, especially the cornea and the lens, are naturally adapted for remote vision in order to see clearly objects situated beyond 15 feet (5 m) from the eye. The lens maintains a relatively flattened shape so that the light rays coming from distant objects will be focused on the retina and will produce a clear image. This does not happen with near vision. If no change in lens shape is made, the image of objects situated a few meters away will be blurred. This does not happen because the eye has a mechanism called accommodation. When one looks at a close object, the ciliary muscle is contracted and the shape of the lens is changed so that the light rays will be precisely diverted to focus perfectly on the retina.

## Refraction defect due to myopia

Normal

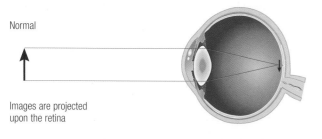

Images are projected upon the retina

Myopia

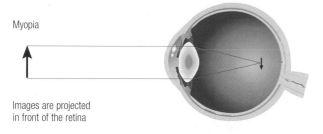

Images are projected in front of the retina

## Optical correction of retina

Incorrect refraction

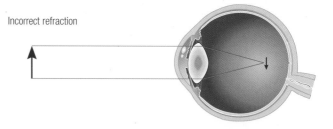

Corrected refraction

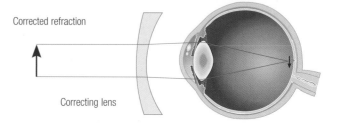

Correcting lens

Myopia is a defect of ocular refraction. The light rays coming from distant objects are focused in front of the retina. Consequently, a blurred vision is perceived. Generally, this is because the eyeball has an anterior to posterior diameter that is greater than normal. The problem can be easily corrected by wearing concave spherical lenses, which diverge the light rays. This type of lens separates the light rays that go through it.

Therefore, if placed in front of the eye, the lens diverts the light rays in such a way that the eye's own refraction elements can focus the rays on the surface of the retina. Today, one can also resort to surgery using lasers to modify the curvature of the cornea and ocular refraction power.

## Refraction defect due to hyperopia

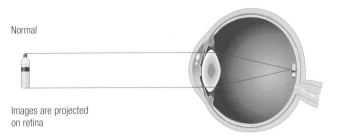

Normal

Images are projected
on retina

Hyperopia

Images are projected
behind retina

Hyperopia is an ocular refraction defect caused by light rays coming from close objects focusing toward a point situated behind the retina. Consequently, blurred near vision is created. This defect is usually due to the fact that the eyeball has an anterior to posterior diameter that is shorter than normal.

## Correction of hyperopia

Incorrect refraction

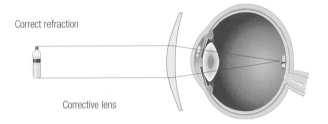

Correct refraction

Corrective lens

The problem is easily corrected by using convex spherical lenses, which converge the light rays. This type of lens brings the light rays that pass through it closer together. Therefore, if placed in front of the eye, the lens diverts the light rays in such a way that the eye can easily focus them upon the surface of the retina.

## Refraction defect due to astigmatism

Normal

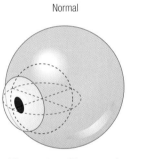

Astigmatism

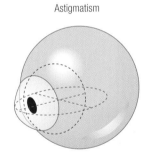

The curvature of the cornea is identical along all meridians.

The curvature of the cornea is different at different meridians.

Normal

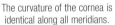

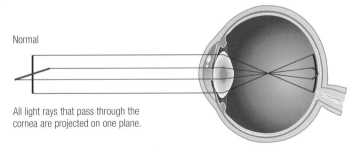

All light rays that pass through the cornea are projected on one plane.

Astigmatism

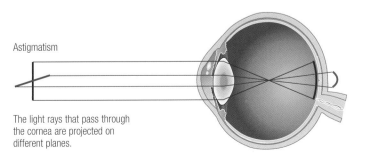

The light rays that pass through the cornea are projected on different planes.

## Optical correction of astigmatism

Incorrect refraction

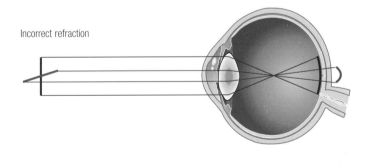

Correct refraction

Corrective lens

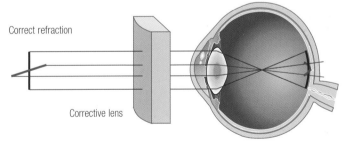

Astigmatism is a defect due to a change in the curvature of the cornea and causes a distorted vision of images. Under normal conditions, the cornea has a hemispherical shape and the curvature of all of its meridians is practically identical. The light rays that pass through this transparent disc, therefore, are concentrated on one plane and permit perfect vision. If there is an unequal curvature among the different meridians, the light rays that pass through the cornea are deviated in such a way that they are projected on different planes. Vision then turns out to be distorted. The problem is corrected by wearing cylindrical lenses that divert the path of the light rays along a particular axis without affecting any other axes.

# Eye and vision problems

**The eye and vision can be affected by widely diverse alterations** that may differ in terms of degree of severity. Some alterations are frequent, while others are extremely rare. All of them have one thing in common. Since sight is the sense that provides the most information about the world surrounding us, its problems generate a variety of repercussions.

## Types of strabismus

Convergent strabismus: The eye is diverted inward in such a way that the gaze seems crossed.

Divergent strabismus: The eye is diverted outward.

Vertical strabismus: The eye is diverted upward or downward.

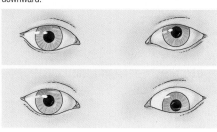

Concomitant strabismus: The angle of eye deviation remains constant in all directions of view.

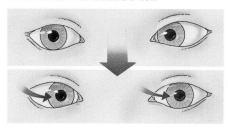

Paralytic strabismus: The angle of deviation of the eyes varies as a function of the direction of the look.

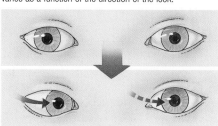

## Strabismus

Strabismus involves the loss of ocular parallelism. The vision axes of both eyes should always be directed at the same object. In strabismus, though, one eye's axis is deviated. The problem is due to a paralysis or to failure of coordination of the extrinsic muscles of the eye that control eye movements and enable the cerebrum to receive complementary images from both eyes. The consequences depend on the age at which this disorder occurs.

When strabismus occurs in adults, it creates double vision. A distinct image is formed in each eye, and the cerebrum cannot merge both of them into a single picture. When it occurs in childhood, it does not cause double vision because the mechanism that allows the cerebrum to merge the images coming from both eyes requires maturation that happens during the first years of life. If the cerebrum receives two very different images, it suppresses one and processes only one. Initially, this is a temporary suppression and both eyes potentially maintain their visual acuity. If strabismus persists, however, the deviated eye loses its vision capacity.

## Extrinsic muscles of the eye

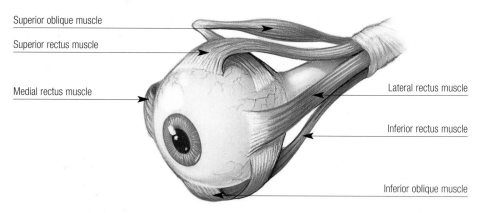

Superior oblique muscle

Superior rectus muscle

Medial rectus muscle

Lateral rectus muscle

Inferior rectus muscle

Inferior oblique muscle

*The movements of each eye depend on the action of six tiny muscles that are inserted in the surface of the eyeball. Perfect coordination between the muscles of both eyes is required so that both eyeballs move in the same direction. For example, for lateral movements, while the medial rectus muscle of an eye is contracted, the lateral rectus muscle is relaxed. The opposite happens in the case of the other eye.*

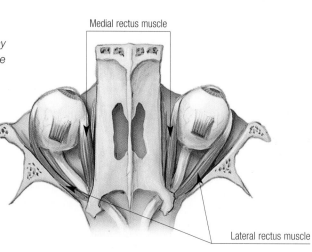

Medial rectus muscle

Lateral rectus muscle

*To correct strabismus, vision exercises are performed to "train" the weakest eye muscles. This treatment, called orthoptic, very often makes it possible to achieve correct ocular parallelism.*

## Mechanism for perceiving colors

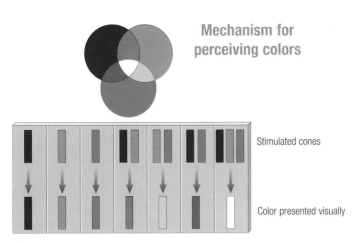

Stimulated cones

Color presented visually

## Tables of colors for the diagnosis of daltonism

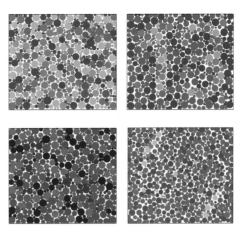

*The following entries must be perceived if a person has correct vision: 182, 13, F4, and 69.*

### Daltonism

Is a congenital hereditary disorder of color vision characterized by the inability to perceive certain colors. Photoreceptors that are sensitive to the colors, the cones, are differentiated into three types, each capable of picking up only one basic color: red, green, or blue. Under normal conditions, the simultaneous and partial stimulation of the three types of cones makes it possible to distinguish a broad range of color views. In daltonism, there is a decline or even a total absence of some of the different types of cones, which is why a person cannot distinguish the colors to which these cones are sensitive. Usually, a person with daltonism cannot detect red or green. To diagnose the problem, one usually employs cards with different colored dots among which there are some of a certain color that form letters or numbers. Persons with normal vision can distinguish these symbols. Persons with a chromatic deficiency will not. They will confuse the colors.

## Types of cataracts

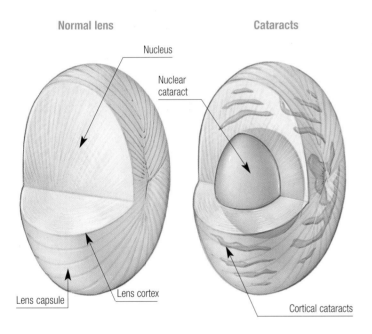

Normal lens

Nucleus

Nuclear cataract

Lens capsule

Lens cortex

Cataracts

Cortical cataracts

### Cataract

A cataract amounts to an opacification of the lens with the subsequent loss of the transparency that characterizes the lens of the eye under normal conditions. Its chief manifestation comes in the form of a diminution of visual acuity, a loss of vision that is more or less accentuated depending on the extent and location of the opaque zone. Any change in arrangement of the elements constituting the lens can lead to the formation of an opaque zone, be it in the central part (nuclear cataract) or in the peripheral zone (cortical cataract) with the subsequent repercussions on vision. Sometimes the problem is congenital. However, the vast majority of cases occur in advanced age, as a result of the changes undergone by the lens with the passage of years, especially as a result of the loss of its aqueous content and the condensation of its fibers. Surgery is the only possible treatment.

## Example of surgical technique used to extirpate the opacified lens.

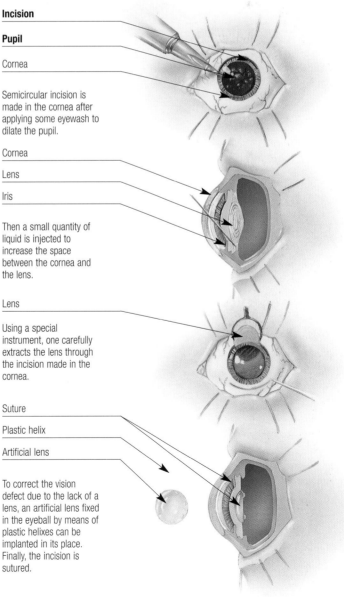

**Incision**

**Pupil**

Cornea

Semicircular incision is made in the cornea after applying some eyewash to dilate the pupil.

Cornea

Lens

Iris

Then a small quantity of liquid is injected to increase the space between the cornea and the lens.

Lens

Using a special instrument, one carefully extracts the lens through the incision made in the cornea.

Suture

Plastic helix

Artificial lens

To correct the vision defect due to the lack of a lens, an artificial lens fixed in the eyeball by means of plastic helixes can be implanted in its place. Finally, the incision is sutured.

# Anatomy of the ear

**The ear is a complex organ that senses sound. Therefore, it is a** basic tool in warning us about what is happening in the environment. It also enables us to communicate with other people. In addition, the ear is involved in maintaining balance.

## Cross section of the ear

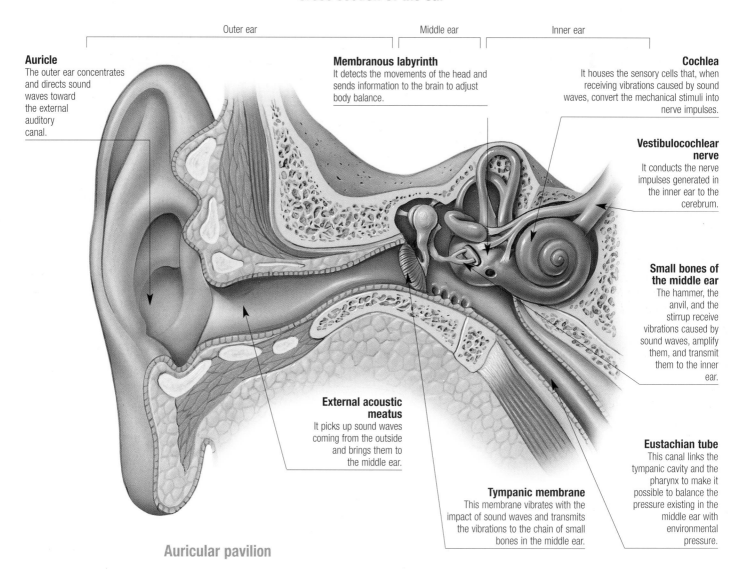

Outer ear    Middle ear    Inner ear

**Auricle**
The outer ear concentrates and directs sound waves toward the external auditory canal.

**Membranous labyrinth**
It detects the movements of the head and sends information to the brain to adjust body balance.

**Cochlea**
It houses the sensory cells that, when receiving vibrations caused by sound waves, convert the mechanical stimuli into nerve impulses.

**Vestibulocochlear nerve**
It conducts the nerve impulses generated in the inner ear to the cerebrum.

**Small bones of the middle ear**
The hammer, the anvil, and the stirrup receive vibrations caused by sound waves, amplify them, and transmit them to the inner ear.

**External acoustic meatus**
It picks up sound waves coming from the outside and brings them to the middle ear.

**Eustachian tube**
This canal links the tympanic cavity and the pharynx to make it possible to balance the pressure existing in the middle ear with environmental pressure.

**Tympanic membrane**
This membrane vibrates with the impact of sound waves and transmits the vibrations to the chain of small bones in the middle ear.

## Auricular pavilion

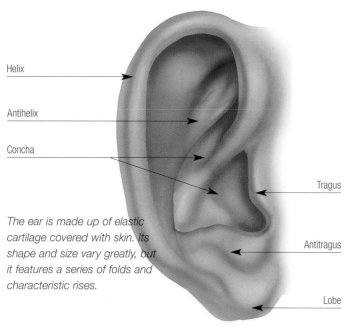

Helix

Antihelix

Concha

Tragus

Antitragus

Lobe

*The ear is made up of elastic cartilage covered with skin. Its shape and size vary greatly, but it features a series of folds and characteristic rises.*

## Eustachian tube

We distinguish three sectors in the ear whose functions differ.

• The **outer ear**, consisting of the ear, the auricle, and the external auditory canal, participates only in hearing.

• The **middle ear**, situated in a cavity of the temporal bone called the tympanic cavity, is separated from the outer ear by a vibrating membrane, the tympanic membrane, sheltering inside it a chain of three little articulated bones that participate only in hearing.

• The **inner ear**, also called the labyrinth, is made up of two parts with different functions. The anterior portion, called the cochlea, is where we find the organ of Corti, which is responsible for hearing. The posterior portion, called the membranous labyrinth, generates stimuli that are involved in maintaining body balance.

## Chain of little bones in the middle ear

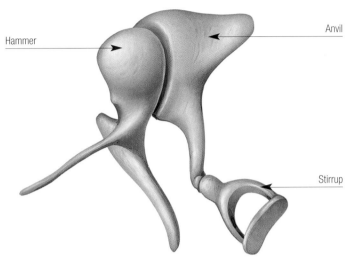

Hammer

Anvil

Stirrup

## Inner ear

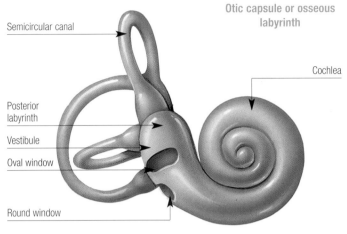

Otic capsule or osseous labyrinth

Semicircular canal

Cochlea

Posterior labyrinth

Vestibule

Oval window

Round window

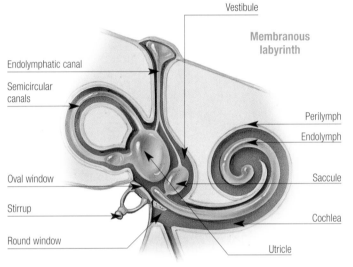

Vestibule

Membranous labyrinth

Endolymphatic canal

Semicircular canals

Perilymph

Endolymph

Oval window

Saccule

Stirrup

Cochlea

Round window

Utricle

The inner ear, or labyrinth, consists of a bony skeleton with a very hard consistency, the otic capsule or osseous labyrinth, within which we find the membranous labyrinth, a structure with a shape that is almost identical but consisting of membranous tissue. The inside of the inner ear is made up of bone, but it is filled with fluid. The fluid, called perilymph, circulates between the osseous labyrinth and the membranous labyrinth, while the inside of the membranous labyrinth is occupied by a fluid called endolymph. The anterior portion, the cochlea, contains the structures that generate the auditory impulses. The rest of the labyrinth, which is involved in the adjustment of body balance, has the saccule containing calcium carbonate particles, the vestibule, and three arc-shaped canals, the semicircular canals, each one arranged in one of the spatial planes.

## Membranous cochlea

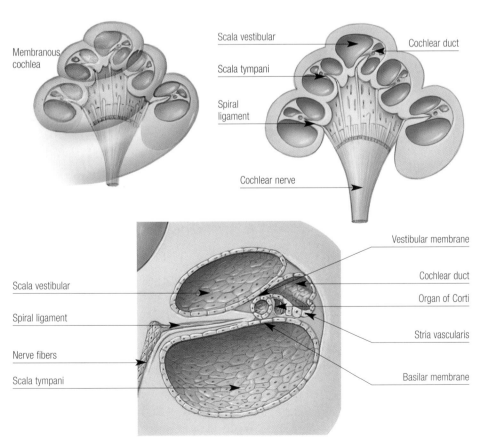

Membranous cochlea

Scala vestibular

Cochlear duct

Scala tympani

Spiral ligament

Cochlear nerve

Scala vestibular

Vestibular membrane

Spiral ligament

Cochlear duct

Organ of Corti

Nerve fibers

Stria vascularis

Scala tympani

Basilar membrane

When seen in cross section, the spiral-shaped cochlea contains a triangular portion—the cochlear duct—containing endolymph. The cochlear duct is surrounded by the scala vestibuli and the scala tympani, which are incompletely separated. The openings of the cochlea are covered by fine membranes that separate the inner ear from the middle ear. The scala vestibuli and the scala tympani shelter these openings. The scala tympani starts at the oval window, and the scala vestibuli extends all the way to the round window. Since it has a triangular shape, the cochlear duct has three faces. The superior face is separated from the scala vestibuli by the vestibular membrane. The inferior face is separated from the scala tympani by the basilar membrane. The lateral face, which adheres to the cochlea, constitutes the stria vascularis, where endolymph is produced. Inside the cochlea is the organ of Corti—the organ specifically responsible for hearing.

# Hearing

**Diverse structures constitute the hearing mechanism. Sound** waves consisting of vibrations of air molecules that expand from the point where the sound is produced are picked up by the outer ear, amplified by the middle ear, and converted by the inner ear into nerve impulses that travel to the cerebrum, where they are perceived.

## Physiology of hearing

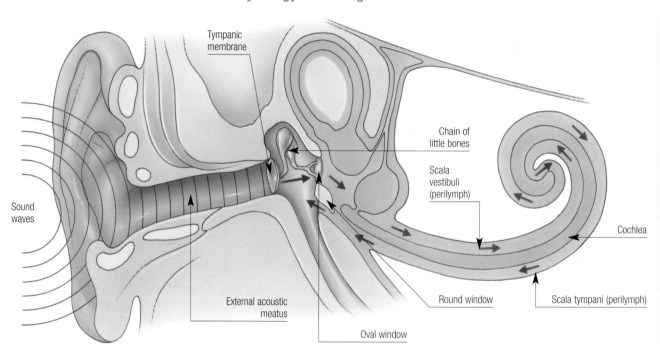

Tympanic membrane

Chain of little bones

Scala vestibuli (perilymph)

Cochlea

Sound waves

Round window

Scala tympani (perilymph)

External acoustic meatus

Oval window

## Hearing mechanism in the inner ear

Sound waves are picked up by the ear and are moved through the external acoustic meatus to the tympanic membrane, which separates the outer ear from the middle ear. The vibrations of the tympanic membrane are transmitted to the chain of little bones in the middle ear, which strike the oval window so that the vibrations may pass through the inner ear that is filled with fluid. When the oval window vibrates, it generates a movement of the perilymph. This produces a kind of "wave" that runs through the entire cochlea, first along the scala vestibuli and then along the scala tympani, until it disappears in the round window. Along its way, the movement of the perilymph causes the stimulation of the organ of Corti, situated in the cochlea, generating the stimuli that are transmitted to the cerebrum via the cochlear nerve.

The movement of the perilymph causes the vibration of the basilar membrane that constitutes the bottom of the organ of Corti. When the sensory cells are moved by the vibrations, the small cilia of its upper surface strike the tectorial membrane and generate metabolic changes that convert the mechanical stimuli into nerve impulses. These are transmitted to the fibers of the cochlear nerve and, through the auditory nerve, extend all the way to the cerebrum, where sound perception is recorded.

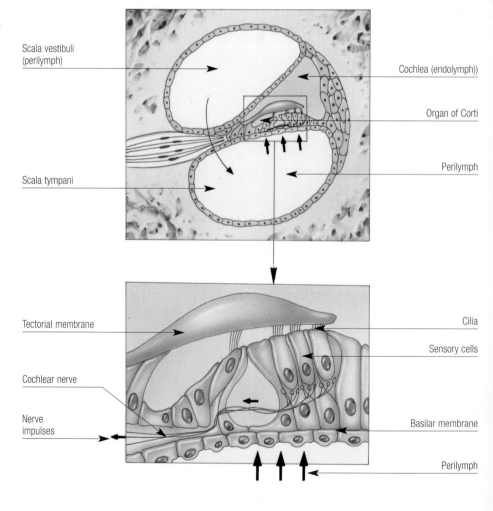

Scala vestibuli (perilymph)

Cochlea (endolymph))

Organ of Corti

Perilymph

Scala tympani

Tectorial membrane

Cilia

Sensory cells

Cochlear nerve

Nerve impulses

Basilar membrane

Perilymph

## Function of chain of little bones in middle ear

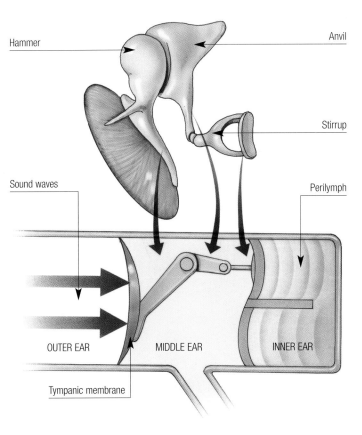

Hammer

Anvil

Stirrup

Sound waves

Perilymph

OUTER EAR    MIDDLE EAR    INNER EAR

Tympanic membrane

## Auditory nerve paths

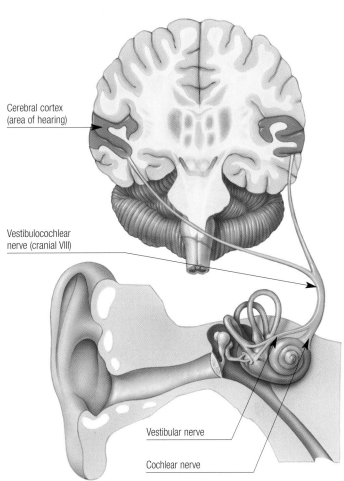

Cerebral cortex (area of hearing)

Vestibulocochlear nerve (cranial VIII)

Vestibular nerve

Cochlear nerve

When the tympanic membrane vibrates, it moves the chain of little bones of the middle ear. Each vibration causes a movement of the hammer. The hammer moves the anvil, and the latter moves the stirrup, whose base strikes the oval window and thus creates a wave in the fluid contained in the inner ear. Since the tympanic membrane has a larger surface than the oval window, sound is concentrated and intensified by the little bones to compensate for the loss of energy by the sound waves as they pass from an air medium to a liquid medium. Even the slightest sounds can be received as a result of this mechanism.

### Acoustic trauma

This is a loss of hearing due to exposure of strong, one-time noises (for example, an explosion) or prolonged noises (music clubs, concerts, work stations). As a result of an acoustic trauma, a person will have normal hearing at the low frequencies (low tones) but will experience a significant reduction in the ability to hear the high frequencies (sharp tones).

Hearing can be protected against acoustic traumas simply by using protective devices and common sense.

The human ear can perceive only sound waves that have certain frequency and intensity. As to frequency, humans can pick up sounds only between 16 and 20,000 Hertz (vibrations per second), with special sensitivity for sounds emitted by the human voice, between 1,000 and 4,000 Hertz. As for intensity, which depends on the amplitude of sound waves, sounds must attain a threshold of 10 decibels, below which they are inaudible.

## Frequency and sound intensity suitable for the human ear

Frequency (hertz)

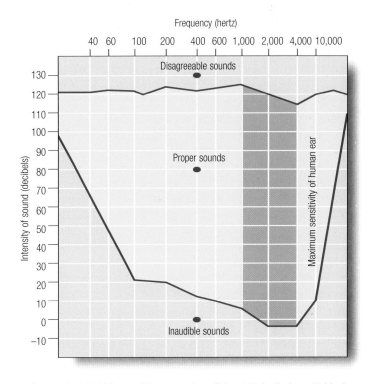

Intensity of sound (decibels)

Maximum sensitivity of human ear

Disagreeable sounds

Proper sounds

Inaudible sounds

*A sound of 400 Hertz will become inaudible at 0 decibels, suitable for the human ear at 80 decibels, and disagreeable at 130 decibels.*

# Balance

**Information supplied by the inner ear to the central nervous** system on the positions and movements of the head is of the utmost importance when it comes to adjusting the state of our musculature to changes automatically so that we can be in perfect equilibrium when we walk.

## Vestibular system

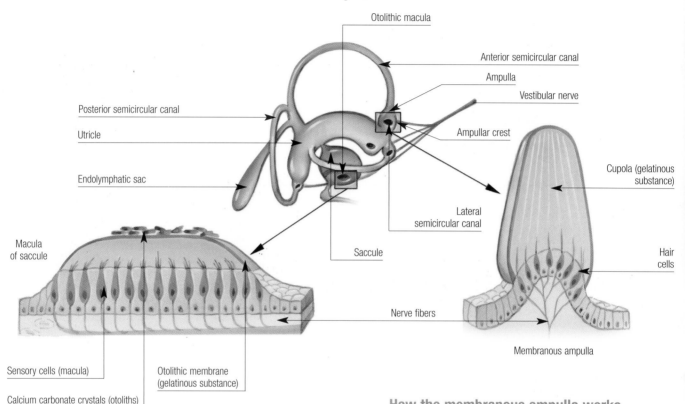

Otolithic macula

Anterior semicircular canal

Ampulla

Vestibular nerve

Posterior semicircular canal

Ampullar crest

Utricle

Cupola (gelatinous substance)

Endolymphatic sac

Lateral semicircular canal

Saccule

Hair cells

Macula of saccule

Nerve fibers

Membranous ampulla

Sensory cells (macula)

Otolithic membrane (gelatinous substance)

Calcium carbonate crystals (otoliths)

## How the vestibular system works

The saccule and the utricle, filled with endolymphatic fluid, contain structures called the macula of saccule. One macula is in the horizontal plane; the other is in the vertical plane. In the maculas are hair cells with ciliated ends. These cilia are immersed in a gelatinous mass that contains tiny calcium carbonate crystals called otoliths. The weight of the otoliths causes the cilia and hair cells to bend as the head moves. Depending on the degree of cellular distortion, nerve stimuli are generated. These stimuli move through the vestibular nerve to the brain, providing information about the position of the head in space and its linear movements.

The semicircular canals that empty in the utricle are arc shaped. Each is in one of the three spatial planes. Each consists of a small dilation at one end, called the ampulla, where there is a membranous ampulla containing sensory cells. The small superficial cilia of these cells are wrapped in a gelatinous mass that, as a result of the movements of the head, is displaced by the endolymph. When performing an angular or rotary movement of the head, sensory cells generate nerve stimuli depending on the degree of torsion of their cilia. This information is sent to the brain. Since each semicircular canal is situated in one spatial plane, by using the information it receives from all of them, the cerebrum can detect the direction and intensity of the rotary movements and the angular accelerations in any spatial plane.

## How the membranous ampulla works

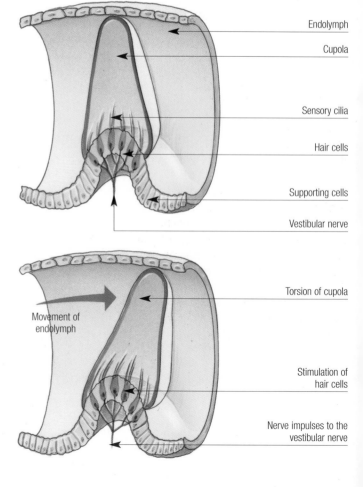

Endolymph

Cupola

Sensory cilia

Hair cells

Supporting cells

Vestibular nerve

Torsion of cupola

Movement of endolymph

Stimulation of hair cells

Nerve impulses to the vestibular nerve

## Elements involved in balance

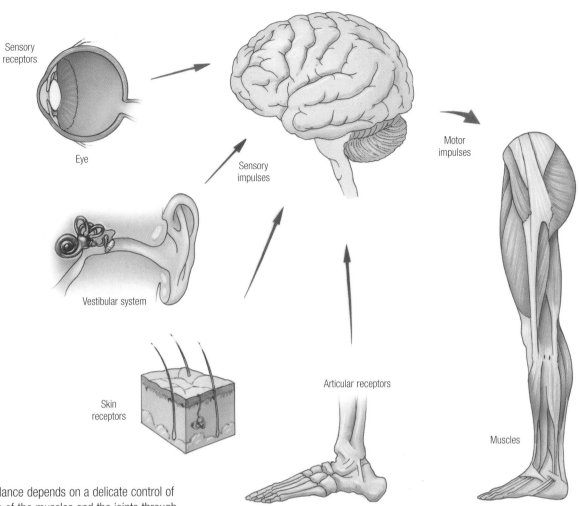

Sensory receptors

Eye

Vestibular system

Skin receptors

Sensory impulses

Motor impulses

Articular receptors

Muscles

Body balance depends on a delicate control of the state of the muscles and the joints through the central nervous system, an unconscious, constant, and dynamic control. In order not to fall, it is necessary at every moment to maintain a certain muscle tension, making sure that some muscles are more contracted and others more relaxed. Pertinent adjustments must be made in response to each and every movement. To arrange for the proper adjustments, the central nervous system must have precise information on the position of every part of the body at every moment.

That information comes from many sources. Sensory receptors located in the skin and joints provide information on the posture of the body in space and on the relative position of all parts of the body. Vision provides an overall idea of the location of the body with respect to the surrounding area and some reference of utmost importance. The vestibular system of the inner ear provides information on the position and movements of the head.

## Transmission of information from the labyrinth to the central nervous system

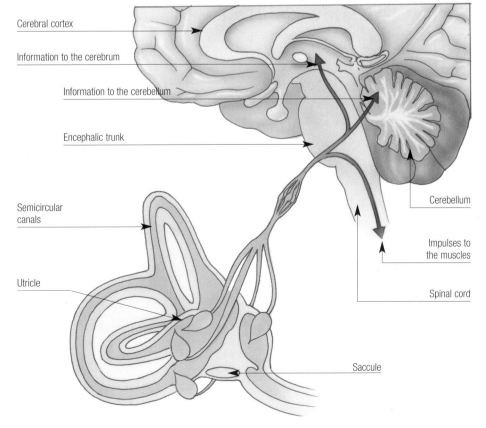

Cerebral cortex

Information to the cerebrum

Information to the cerebellum

Encephalic trunk

Semicircular canals

Utricle

Cerebellum

Impulses to the muscles

Spinal cord

Saccule

# Anatomy of the skin

**The skin is a thick, strong, and flexible membrane composed** of three layers: epidermis, dermis, and hypodermis. It also has various attached structures: the sweat glands, the sebaceous glands, the sensory receptors, the hair follicles, and the nails. The skin not only serves as a body covering but also performs other important functions.

## Three-dimensional representation of piece of skin

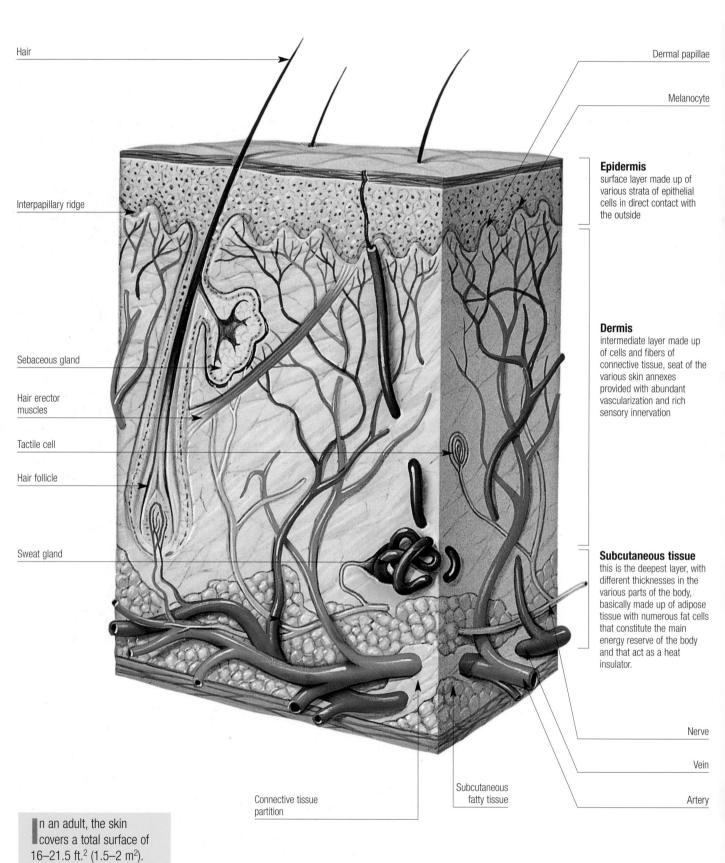

Hair

Interpapillary ridge

Sebaceous gland

Hair erector muscles

Tactile cell

Hair follicle

Sweat gland

Connective tissue partition

Subcutaneous fatty tissue

Dermal papillae

Melanocyte

**Epidermis**
surface layer made up of various strata of epithelial cells in direct contact with the outside

**Dermis**
intermediate layer made up of cells and fibers of connective tissue, seat of the various skin annexes provided with abundant vascularization and rich sensory innervation

**Subcutaneous tissue**
this is the deepest layer, with different thicknesses in the various parts of the body, basically made up of adipose tissue with numerous fat cells that constitute the main energy reserve of the body and that act as a heat insulator.

Nerve

Vein

Artery

In an adult, the skin covers a total surface of 16–21.5 ft.$^2$ (1.5–2 m$^2$).

## Microscopic structure of the skin

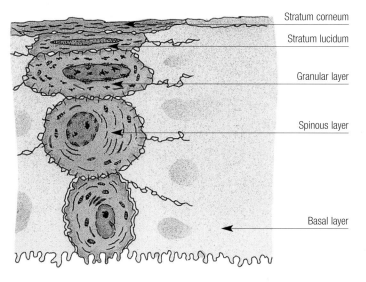

Stratum corneum
Stratum lucidum
Granular layer
Spinous layer
Basal layer

The epidermis, the layer of the skin that is directly in contact with the outside, has a thickness that fluctuates in various parts of the body between 0.05 and 0.5 mm. It is made up of epithelial tissue consisting of only back-to-back cells that do not have any intercellular substance between them. These cells are arranged in superimposed strata that make up four or five layers, depending on the sector of the body.

### Regeneration of the epidermis

The epidermis is constantly renewed because the surface cells, exposed to the outside environment, are continuously shed and are then replaced by others from underneath. Therefore, cells of the basal layer are multiplied continuously and new ones push those that are further up toward the surface, crossing the various layers as they are modified and lose vitality, until reaching the stratum corneum. After a certain period of time, these fall away. This process takes between 20 and 30 days.

## Troughs and crests of the skin

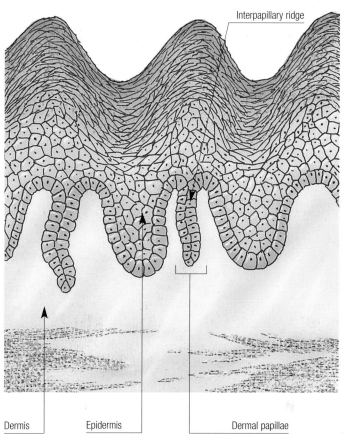

Interpapillary ridge

Dermis | Epidermis | Dermal papillae

The dermis, which is under the epidermis and is separated from the former by a thin basal membrane, features numerous folds. We can detect some conical protrusions of the dermis that are projected toward the epidermis. In other words, the dermal papillae, interspersed with protrusions on the epidermis, are projected toward the dermis, the interpapillary ridges. This considerably increases the contact surface between both layers. This contact is very important because nutrition of the epidermis depends on the blood vessels that reach only to the dermis.

## Melanin

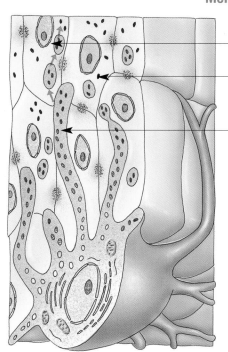

Cells of epidermis
Granules of melanin
Granules of melanin in the process of forming

Interspersed between the epithelial cells in the depth of the epidermis are cells responsible for synthesizing melanin, a dark pigment whose concentration gives each skin a coloration of its own and that is also responsible for the phenomenon of tanning: the melanocytes. These cells, which are round and have numerous prolongations, contain tiny granules called melanosomes in which melanin is produced due to the influence of hormonal factors and the ultraviolet radiation of the sun.

### Skin coloration

The color of the skin depends on two factors. The blood that circulates through the network of capillaries of the dermis and becomes transparent on the skin surface gives skin a pink tone. The color also depends on the content, quantity, and distribution of melanin, a pigment that has the function of absorbing the rays of the sun and preventing their passage inside the body, where they exert harmful effects. The production of the pigment through melanocytes is regulated by genetic and hormonal factors. This explains the diverse skin coloration of individuals of different races and of each person, in particular. The chief stimulus for the production of melanin is exposure to the sun, which causes tanning.

# Functions of the skin

**The skin acts as a barrier that protects the surface of the body** against potentially aggressive agents from the external environment and prevents their passage into the interior of the body. It also performs other relevant functions, such as adjusting body temperature.

## Involuntary skin mechanisms to regulate body temperature

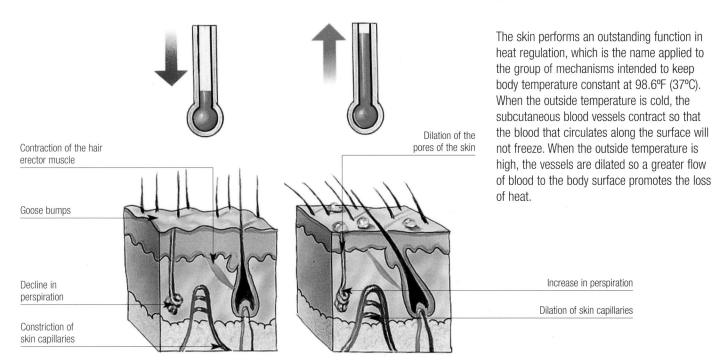

Contraction of the hair erector muscle

Goose bumps

Decline in perspiration

Constriction of skin capillaries

Dilation of the pores of the skin

Increase in perspiration

Dilation of skin capillaries

The skin performs an outstanding function in heat regulation, which is the name applied to the group of mechanisms intended to keep body temperature constant at 98.6ºF (37ºC). When the outside temperature is cold, the subcutaneous blood vessels contract so that the blood that circulates along the surface will not freeze. When the outside temperature is high, the vessels are dilated so a greater flow of blood to the body surface promotes the loss of heat.

## Sweat glands

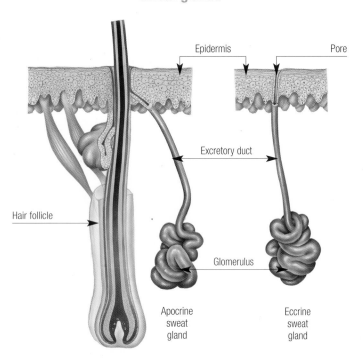

Epidermis

Pore

Excretory duct

Hair follicle

Glomerulus

Apocrine sweat gland

Eccrine sweat gland

## Distribution of sweat glands in the body

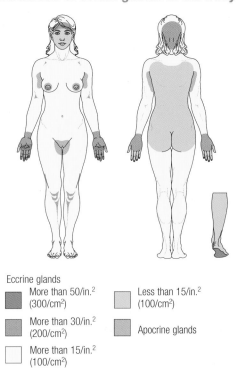

**Eccrine glands**

More than 50/in.² (300/cm²)

More than 30/in.² (200/cm²)

More than 15/in.² (100/cm²)

Less than 15/in.² (100/cm²)

Apocrine glands

### Production of sweat

Sweat is the secretion of the sweat glands. It is made up of water and small quantities of salts and various chemical substances derived from metabolism. There are two types of sweat glands. The eccrine glands, which are the most abundant, empty into some tiny pores on the surface of the skin. The apocrine glands discharge their secretions into a hair follicle.

The activity of the sweat glands, controlled by the autonomous nervous system, contributes to regulating body temperature since the evaporation of sweat has a cooling effect on the skin. Every day, at least a pint of sweat is produced. It is barely perceived, but this quantity can increase considerably in a warm environment and as a result of physical exercise.

## Sebaceous gland

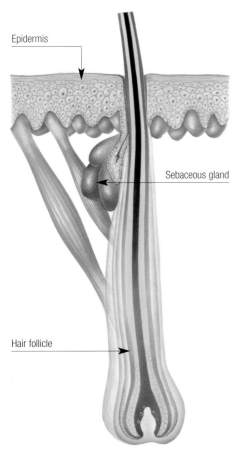

Epidermis

Sebaceous gland

Hair follicle

## Body zones with abundant sebaceous glands

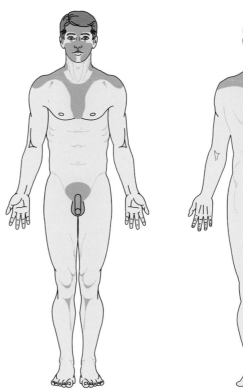

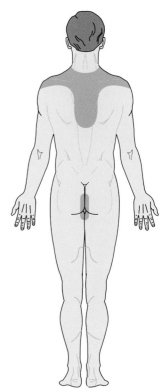

## Function of the sebaceous glands

The sebaceous glands, distributed all over the body surface but more abundant in the face, chest, back, and genital area, process a fatty secretion—sebum—that forms a protective film on the epidermis and lubricates the hair. The sebum has a protective function. It is mixed with the product from the shedding of the epidermis and with sweat, constituting an acid-fat mantle, which, among other things, makes it difficult for germs to form on the skin surface. When the ambient temperature is low, sebaceous secretion is more solidified and obstructs the evaporation of sweat. The result is that it helps maintain body temperature.

## Wound healing

The renewal of skin tissue facilitates prompt and effective repair of wounds, although the consequences depend on the depth of the wound. If only the epidermis is involved, as happens when there is a simple scrape, tissue is regenerated from the basal layer and no visible mark is left. On the other hand, when the dermis is also involved, as usually happens from a cut, the edges of the wound become separated, starting the process of cicatrization.

Granular tissue now proliferates from the edges. It is made up of cells and connective fibers that gradually fill up the empty areas and restore the continuity of the epidermis, which in the end covers the lesion. Since the epidermal layer is thinner than normal and since the connective tissue that repairs the wound does not have the same structure as the original dermis, what is left in this zone is at first a pink and then a whitish mark called a scar.

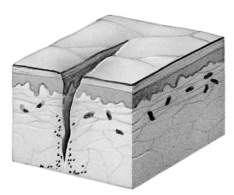

The edges of the wound remain separated

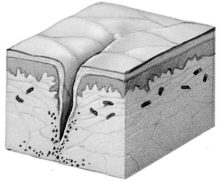

A fibrous tissue, which tends to fill the empty area, proliferates from the edges

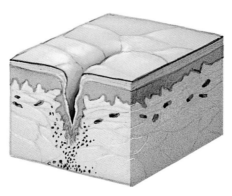

Continuity of the epidermis is restored at the bottom of the lesion

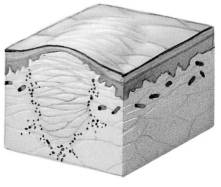

The fibrous tissue pushes the epidermis to the surface

# Hair and nails

**The hair consists of some thin and strong filaments that are** distributed over almost the entire body surface. The nails are thin, semitransparent plates that are hard and resistant. They protect the dorsal face of the last phalanx of the fingers of the hand and the toes of the feet. They are the principal cutaneous annexes.

## Longitudinal and transverse cross section of a hair follicle

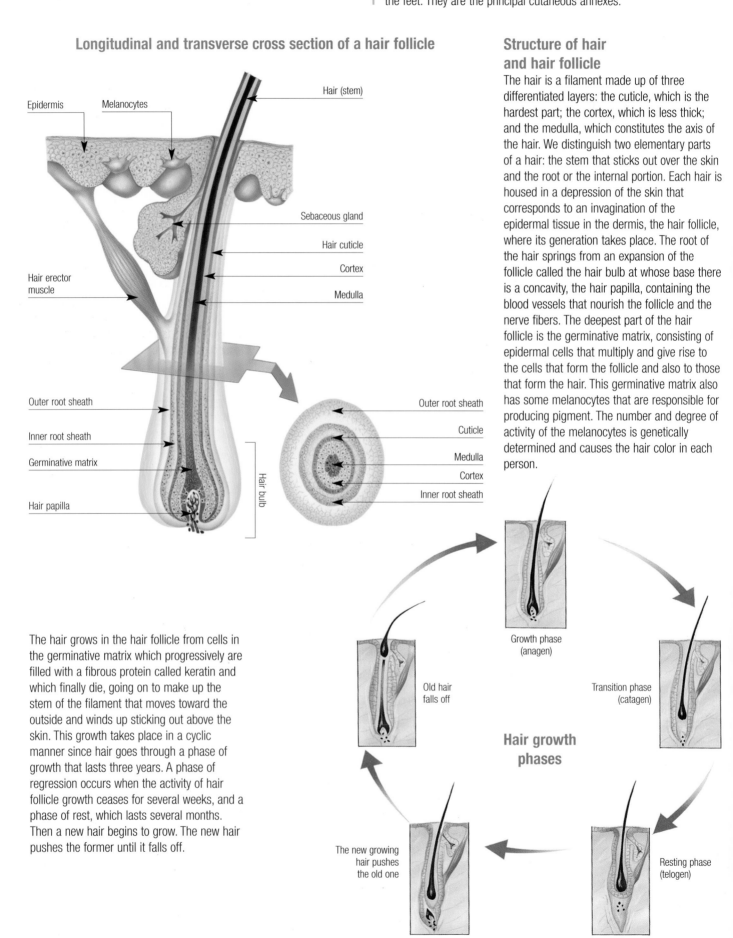

Epidermis

Melanocytes

Hair (stem)

Sebaceous gland

Hair cuticle

Cortex

Medulla

Hair erector muscle

Outer root sheath

Inner root sheath

Germinative matrix

Hair papilla

Hair bulb

Outer root sheath

Cuticle

Medulla

Cortex

Inner root sheath

## Structure of hair and hair follicle

The hair is a filament made up of three differentiated layers: the cuticle, which is the hardest part; the cortex, which is less thick; and the medulla, which constitutes the axis of the hair. We distinguish two elementary parts of a hair: the stem that sticks out over the skin and the root or the internal portion. Each hair is housed in a depression of the skin that corresponds to an invagination of the epidermal tissue in the dermis, the hair follicle, where its generation takes place. The root of the hair springs from an expansion of the follicle called the hair bulb at whose base there is a concavity, the hair papilla, containing the blood vessels that nourish the follicle and the nerve fibers. The deepest part of the hair follicle is the germinative matrix, consisting of epidermal cells that multiply and give rise to the cells that form the follicle and also to those that form the hair. This germinative matrix also has some melanocytes that are responsible for producing pigment. The number and degree of activity of the melanocytes is genetically determined and causes the hair color in each person.

The hair grows in the hair follicle from cells in the germinative matrix which progressively are filled with a fibrous protein called keratin and which finally die, going on to make up the stem of the filament that moves toward the outside and winds up sticking out above the skin. This growth takes place in a cyclic manner since hair goes through a phase of growth that lasts three years. A phase of regression occurs when the activity of hair follicle growth ceases for several weeks, and a phase of rest, which lasts several months. Then a new hair begins to grow. The new hair pushes the former until it falls off.

Growth phase (anagen)

Old hair falls off

Transition phase (catagen)

**Hair growth phases**

The new growing hair pushes the old one

Resting phase (telogen)

## Gray hair–white hair

Hair color depends primarily on how much melanin the hair contains. Melanin is the pigment produced by the melanocytes of the germinative matrix. To a lesser degree, hair color also depends on the arrangement of cells that make up the hair. Round hair, oval hair, and elliptical hair all reflect light differently. Therefore, the shape of each hair affects the color's brilliance and tone.

Over time, the melanocytes of the germinative matrix stop producing pigment. The age when they cease their activity is highly variable. However, the result is typically either gray or white hair.

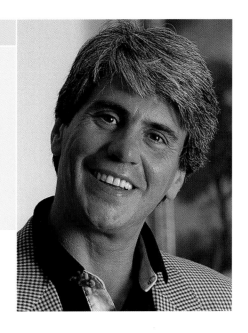

## Development of common alopecia (baldness)

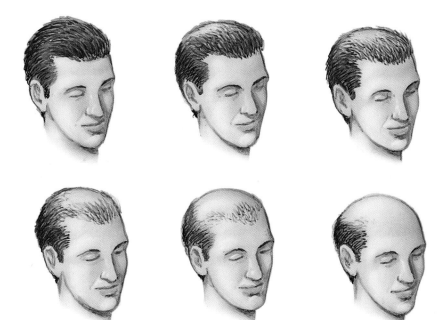

Hair loss is called alopecia or baldness. This phenomenon can be either total or partial and is due to various causes. The most common form is ordinary or androgenic alopecia, which is inherently a male trait. It originates from hereditary genetic factors that cause a particular sensitivity of the hair follicles to the action of the androgens, the male sex hormones. At a certain time in a person's life, the rate of normal hair regeneration is accelerated due to the stimulus provided by these hormones. Hair breaks more easily and turns ever finer as the hair follicles atrophy. In some areas on the head, usually following a well-defined pattern, normal hair growth is gradually replaced by ever-finer hair until it turns into a thin veil that is practically unnoticeable.

## Structure of a nail

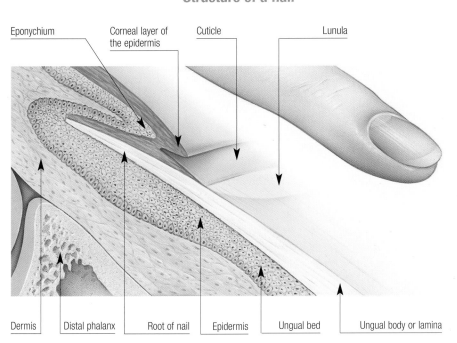

Eponychium | Corneal layer of the epidermis | Cuticle | Lunula

Dermis | Distal phalanx | Root of nail | Epidermis | Ungual bed | Ungual body or lamina

The nails are thin but hard and resistant lamina whose function is to protect the last phalanx of the fingers of the hand and the toes. They are also useful in performing actions that require a certain degree of precision, such as grabbing, folding, or separating. Their structure resembles that of hair since they are basically made of keratin and are produced by the epidermis. Growth begins at the root, which remains hidden from sight, where cells of the corneal and epidermal layer process a very hard keratin that slips over the ungual bed, forming a laminar corresponding to the nail. Although the speed of nail growth varies according to the individual, it usually is about 0.1 mm per day.

# Touch

The skin is the organ of touch, a sense that provides valuable information about the world that surrounds us. It enables us to perceive rubbing and pressure; to identify shape, texture, and other palpable qualities of objects; as well as to distinguish heat variations and provide an alert against external factors that cause painful stimuli.

Distributed over the skin surface are countless receptors that respond to diverse stimuli. Through the sensory paths, they send information to the central nervous system for interpretation. The free ends of the sensory nerves function as these receptors. They perceive tactile and painful stimuli.

Specialized receptors perceive different stimuli. The Vater-Pacini corpuscles, situated deep in the dermis, primarily detect vibrations and changes in pressure that take place on the skin. Meissner's corpuscles, located in the dermal papillae, respond to tactile stimuli and are especially abundant in the fingertips and lips. Krause's end bulbs, located in the superficial area of the dermis, are sensitive to cold. Ruffini's corpuscles, located deep in the dermis, are sensitive to heat.

## Sensory receptors of the skin

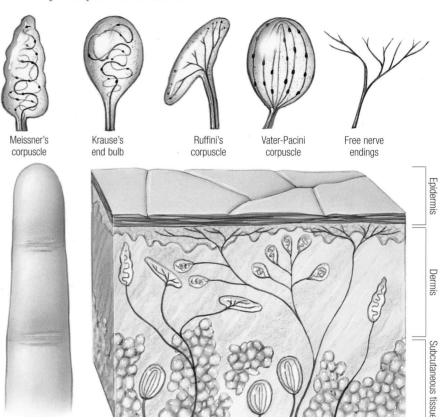

Meissner's corpuscle

Krause's end bulb

Ruffini's corpuscle

Vater-Pacini corpuscle

Free nerve endings

Epidermis

Dermis

Subcutaneous tissue

## Sensory nerve paths

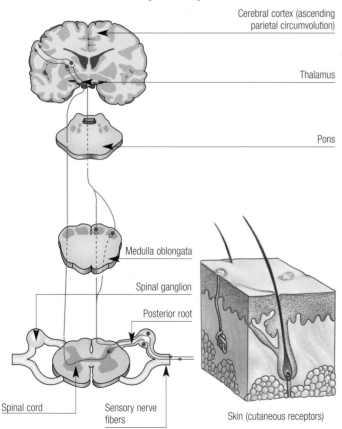

Cerebral cortex (ascending parietal circumvolution)

Thalamus

Pons

Medulla oblongata

Spinal ganglion

Posterior root

Spinal cord

Sensory nerve fibers

Skin (cutaneous receptors)

*The capacity for discriminating tactile stimuli can be broadly developed through practice. This faculty is manifest in the case of blind persons who learn to read with the Braille system based on an alphabet of raised dots that can be interpreted by sliding the fingers over them.*

Mechanical, heat, and pain stimuli detected by cutaneous receptors pass through sensory nerves that extend all the way to the medulla oblongata and follow an ascending path along nerve fibers that constitute specific cords depending on the type of sensitivity they transmit. In that way, information reaches the brain. After some stops, primarily in the thalamus, the information reaches as far as the cerebrum, specifically, the ascending parietal circumvolution of the cerebral cortex, where the stimuli are interpreted and sensations are perceived.

## Development of the sense of touch

The sense of touch is proportionally the most highly developed in the newborn. It gives the baby information concerning the surrounding world. This is the reason why the baby responds so well to caresses, why the baby acts in a placid manner when it is dry and in a warm environment, and why the baby cries to announce that it is wet. One must basically keep in mind that for the small child, touch is important. Special emphasis on making pleasurable contact with the baby should be made. This treatment will condition the way in which the baby perceives the world.

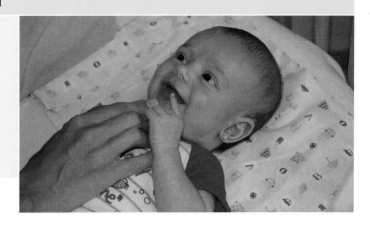

### Reflex action to a painful stimulus

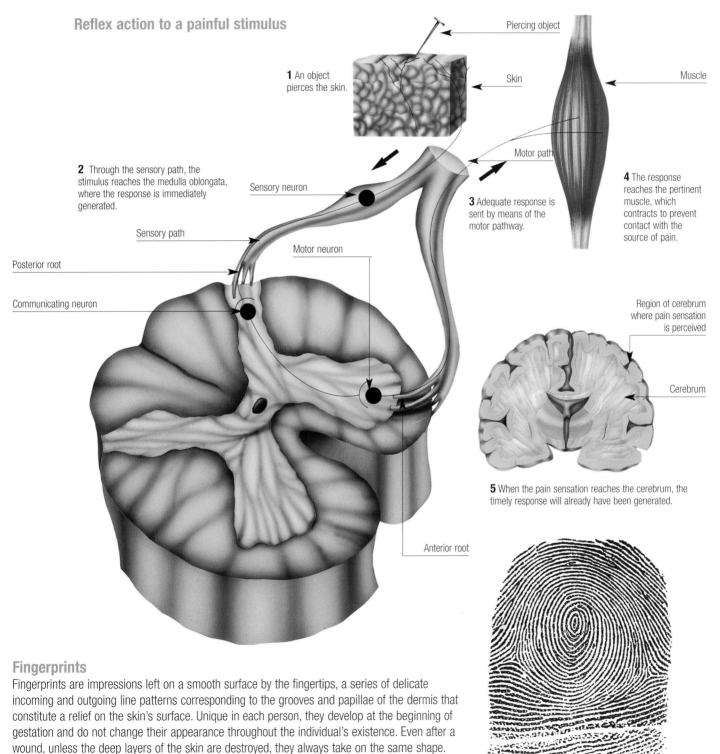

Piercing object

**1** An object pierces the skin.

Skin

Muscle

Motor path

**2** Through the sensory path, the stimulus reaches the medulla oblongata, where the response is immediately generated.

Sensory neuron

**3** Adequate response is sent by means of the motor pathway.

**4** The response reaches the pertinent muscle, which contracts to prevent contact with the source of pain.

Sensory path

Motor neuron

Posterior root

Communicating neuron

Region of cerebrum where pain sensation is perceived

Cerebrum

**5** When the pain sensation reaches the cerebrum, the timely response will already have been generated.

Anterior root

## Fingerprints

Fingerprints are impressions left on a smooth surface by the fingertips, a series of delicate incoming and outgoing line patterns corresponding to the grooves and papillae of the dermis that constitute a relief on the skin's surface. Unique in each person, they develop at the beginning of gestation and do not change their appearance throughout the individual's existence. Even after a wound, unless the deep layers of the skin are destroyed, they always take on the same shape.

# Flavors and scents

function in the digestive process since both an agreeable aroma and a pleasant taste will stimulate salivary and gastric secretions. Smell also provides valuable information about the presence of toxic gases that smell bad.

## The gustatory papillae of the tongue

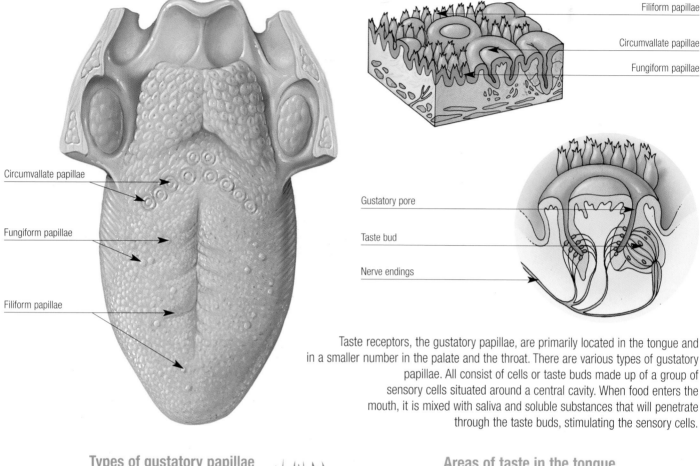

Circumvallate papillae

Fungiform papillae

Filiform papillae

Filiform papillae

Circumvallate papillae

Fungiform papillae

Gustatory pore

Taste bud

Nerve endings

Taste receptors, the gustatory papillae, are primarily located in the tongue and in a smaller number in the palate and the throat. There are various types of gustatory papillae. All consist of cells or taste buds made up of a group of sensory cells situated around a central cavity. When food enters the mouth, it is mixed with saliva and soluble substances that will penetrate through the taste buds, stimulating the sensory cells.

## Types of gustatory papillae

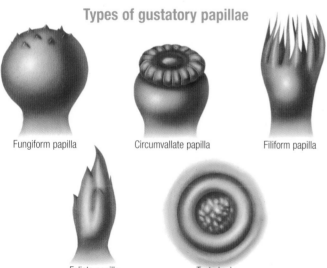

Fungiform papilla

Circumvallate papilla

Filiform papilla

Foliate papilla

Taste bud

All types of gustatory papillae located on the surface of the tongue are capable of perceiving the four basic sensations: sweet, bitter, salty, and sour. Some will react with greater intensity to the various stimuli. We differentiate areas that specialize in the perception of various tastes: sweet at the tip, bitter in the back, sour along the sides, and salty in the front except for the tip.

## Areas of taste in the tongue

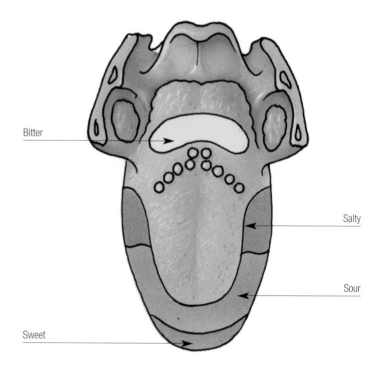

Bitter

Salty

Sour

Sweet

## Gustatory paths

Cerebral cortex

Thalamus

Medulla oblongata

Trigeminal and facial nerves

Glossopharyngeal nerve

Vagus nerve

The stimuli produced in the gustatory papillae depart through the nerve ends of the sensory cells and travel through various nerves that innervate the mouth. From the encephalic trunk, they pass through other specific nerve paths up to the thalamus. When they arrive at the taste area, situated in the parietal lobule of the cerebral cortex, they are decoded and their sensations are perceived. Although there are only four basic taste sensations, the cerebrum interprets infinite patterns depending on the combination of the various tastes and even the smell.

## Mechanism of smell

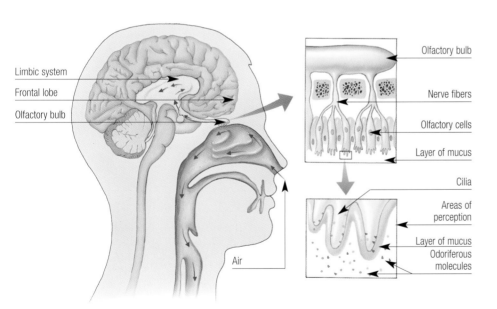

Limbic system

Frontal lobe

Olfactory bulb

Air

Olfactory bulb

Nerve fibers

Olfactory cells

Layer of mucus

Cilia

Areas of perception

Layer of mucus
Odoriferous molecules

## Olfactory cell

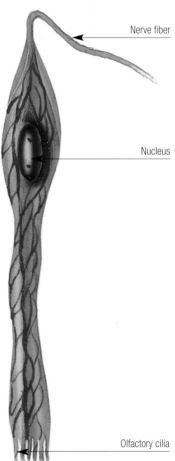

Nerve fiber

Nucleus

Olfactory cilia

The receptors of this sense are located in the olfactory membrane, a small area in the roof of the nasal fossa where there is a layer of cells specializing in the detection of odors. These cells are elongated. At their free end, they have tiny olfactory cilia that are immersed in a layer of mucus generated by the glands in the nasal wall. The volatile molecules that are present in the air we inhale are first dissolved in mucus, link with the reception areas of these cilia, and then generate nerve stimuli in the cells. At the other end, the olfactory cells have thin nerve fibers that pass through the roof of the nasal fossa and extend all the way to the olfactory bulb, from which springs the olfactory nerve that carries the information to the olfactory centers of the cerebral cortex.

# Structure of the nervous system

**The nervous system is made up of an intricate network of** specialized cells that constitute various interrelated structures. These cells control all of the activities of the body, both the voluntary ones as well as those that take place automatically. The nervous system lets us relate to the outside world and is responsible for intellectual functions.

## Components of the nervous system

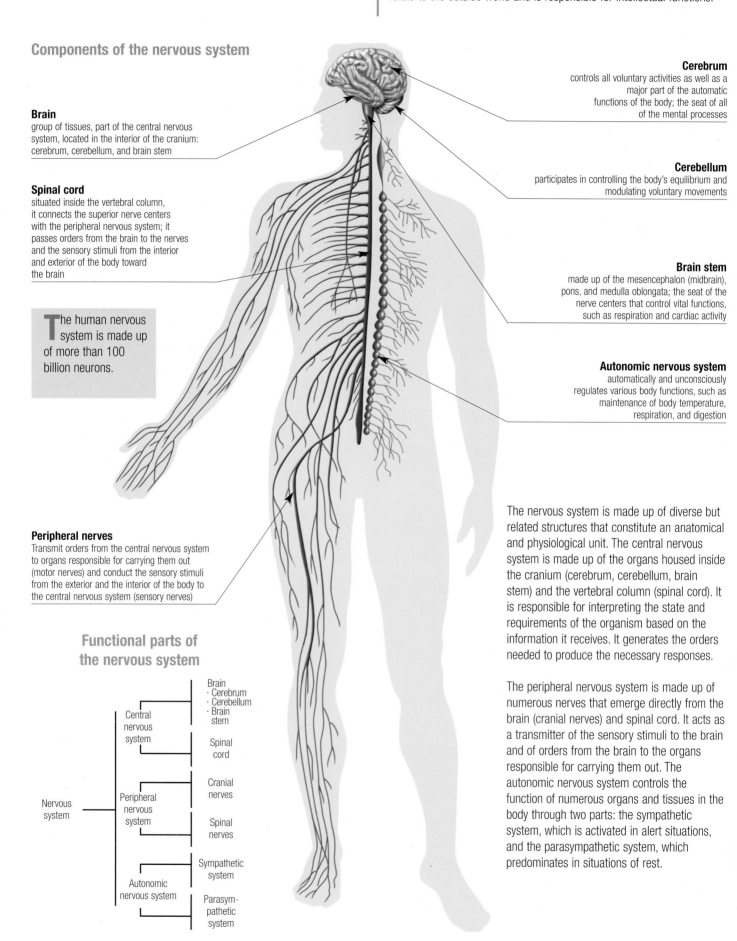

**Cerebrum**
controls all voluntary activities as well as a major part of the automatic functions of the body; the seat of all of the mental processes

**Brain**
group of tissues, part of the central nervous system, located in the interior of the cranium: cerebrum, cerebellum, and brain stem

**Cerebellum**
participates in controlling the body's equilibrium and modulating voluntary movements

**Spinal cord**
situated inside the vertebral column, it connects the superior nerve centers with the peripheral nervous system; it passes orders from the brain to the nerves and the sensory stimuli from the interior and exterior of the body toward the brain

**Brain stem**
made up of the mesencephalon (midbrain), pons, and medulla oblongata; the seat of the nerve centers that control vital functions, such as respiration and cardiac activity

The human nervous system is made up of more than 100 billion neurons.

**Autonomic nervous system**
automatically and unconsciously regulates various body functions, such as maintenance of body temperature, respiration, and digestion

**Peripheral nerves**
Transmit orders from the central nervous system to organs responsible for carrying them out (motor nerves) and conduct the sensory stimuli from the exterior and the interior of the body to the central nervous system (sensory nerves)

The nervous system is made up of diverse but related structures that constitute an anatomical and physiological unit. The central nervous system is made up of the organs housed inside the cranium (cerebrum, cerebellum, brain stem) and the vertebral column (spinal cord). It is responsible for interpreting the state and requirements of the organism based on the information it receives. It generates the orders needed to produce the necessary responses.

## Functional parts of the nervous system

Nervous system
- Central nervous system
  - Brain
    - · Cerebrum
    - · Cerebellum
    - · Brain stem
  - Spinal cord
- Peripheral nervous system
  - Cranial nerves
  - Spinal nerves
- Autonomic nervous system
  - Sympathetic system
  - Parasympathetic system

The peripheral nervous system is made up of numerous nerves that emerge directly from the brain (cranial nerves) and spinal cord. It acts as a transmitter of the sensory stimuli to the brain and of orders from the brain to the organs responsible for carrying them out. The autonomic nervous system controls the function of numerous organs and tissues in the body through two parts: the sympathetic system, which is activated in alert situations, and the parasympathetic system, which predominates in situations of rest.

## Structure of the neuron

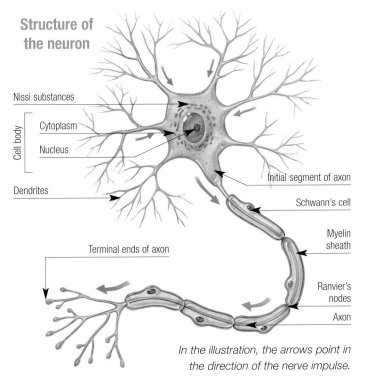

- Nissi substances
- Cell body
  - Cytoplasm
  - Nucleus
- Dendrites
- Initial segment of axon
- Schwann's cell
- Terminal ends of axon
- Myelin sheath
- Ranvier's nodes
- Axon

*In the illustration, the arrows point in the direction of the nerve impulse.*

## Synapse

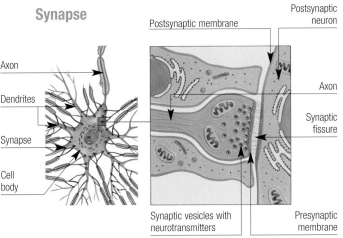
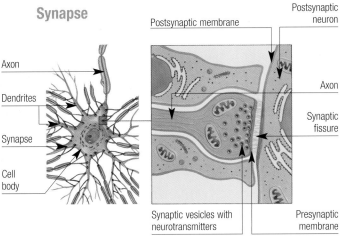

- Axon
- Dendrites
- Synapse
- Cell body
- Postsynaptic membrane
- Postsynaptic neuron
- Axon
- Synaptic fissure
- Synaptic vesicles with neurotransmitters
- Presynaptic membrane

## Types of neurons

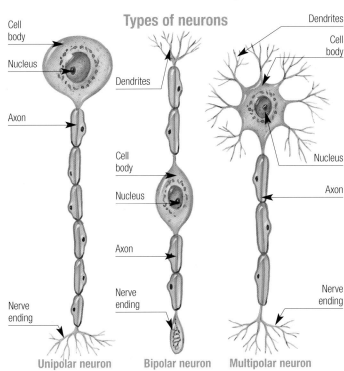

- Cell body
- Nucleus
- Dendrites
- Axon
- Cell body
- Nucleus
- Axon
- Nerve ending
- **Unipolar neuron**
- Nerve ending
- **Bipolar neuron**
- Dendrites
- Cell body
- Nucleus
- Axon
- Nerve ending
- **Multipolar neuron**

## Neuron: the nerve cell

All structures of the nervous system are composed of the same two cell types: neurons and neuroglia. The neurons are responsible for generating and transmitting nerve impulses. The neuroglia provide support, nutrition, and protection to the neurons. The body contains millions of neurons that vary in type, shape, and size. However, they all have the same fundamental structure. Each neuron has a cell body from which springs a receptor portion and a transmitter portion: dendrites and axon. The dendrites are short, treelike branches that receive stimuli from other nerve cells. The axon can vary in length but always ends in tiny branches called nerve endings. The axon transmits the nerve impulse to other nerve cells.

Neurons communicate with each other through signals transmitted by a complex physical-chemical mechanism in the form of nerve impulses. In response to certain stimuli, biochemical changes are produced in the neuron. These changes trigger an electrical signal that runs through the cell along the axon at whose end communication is established with the adjacent neurons. The nerve impulse is not transmitted to the adjacent neurons by direct contact but, rather, through a special connection called the **synapse**. The branches of the axon terminate very close to the adjacent neurons but are always separated by a narrow space, the synaptic cleft. The nerve impulse crosses that space by means of chemical substances called neurotransmitters.

Each neuron processes a specific neurotransmitter that is stored in synaptic vesicles accumulated in the branches of the axon. These vesicles release their content to the synaptic cleft in response to the arrival of an electrical impulse at the axon. Through this space, the neurotransmitter is combined with receptors present on the surface of the adjacent neurons. This generates biochemical changes in the recipient neuron's membrane. The effect depends on the type of neurotransmitter. It can trigger an electrical signal (excitatory synapse), or it can reduce the excitability (inhibiting synapse).

## Gray matter and white matter

The axon of most neurons is covered by an envelope made of a series of concentric layers of a white fatty substance with insulating properties. It is very important in the direct transmission of nerve impulses. It is called the myelin sheath and is composed of some special cells, oligodendrocytes or Schwann cells. The organs of the central nervous system contain areas made up primarily of neuronal bodies, while others contain bundles of nerve fibers corresponding to cellular prolongations called axons. In the first case, we speak of gray matter because that is the predominant color of the neuronal bodies. On the other hand, clusters of nerve fibers, each surrounded by a sheath of whitish myelin, constitute the so-called white matter.

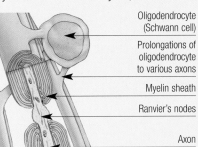

- Oligodendrocyte (Schwann cell)
- Prolongations of oligodendrocyte to various axons
- Myelin sheath
- Ranvier's nodes
- Axon

# Brain

**The brain is the part of the central nervous system that is made** up of the tissues contained inside the cranium. It is surrounded by protective membranes, the meninges, among which circulates a fluid intended to soften traumas. This is called the cerebrospinal fluid and accumulates in dilations called the cerebral ventricles.

## Sagittal cross section of brain

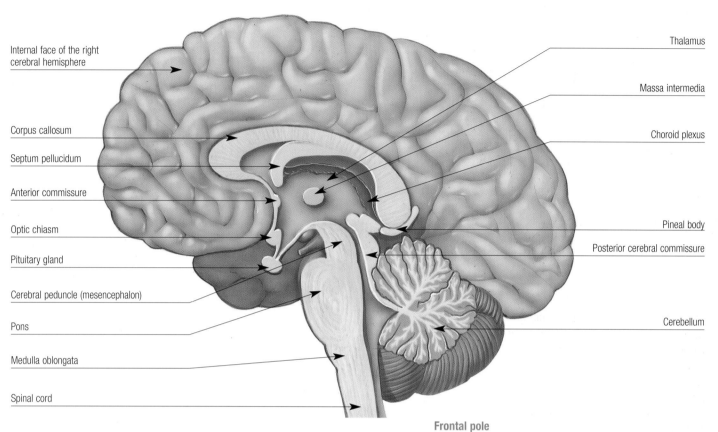

Internal face of the right cerebral hemisphere

Corpus callosum

Septum pellucidum

Anterior commissure

Optic chiasm

Pituitary gland

Cerebral peduncle (mesencephalon)

Pons

Medulla oblongata

Spinal cord

Thalamus

Massa intermedia

Choroid plexus

Pineal body

Posterior cerebral commissure

Cerebellum

## Interior view of brain

The brain consists of the following.
• The **cerebrum** is the most important and voluminous portion. It controls all the voluntary activity and a large part of the involuntary activity of the human body. It is also the seat of mental processes, such as memory and intelligence.
• The **brain stem** is made up of the mesencephalon (midbrain), the pons, and the medulla oblongata, which regulate vital functions. It is also where most cranial nerves originate.
• The **cerebellum** participates in the control of equilibrium and the coordination of body movements.

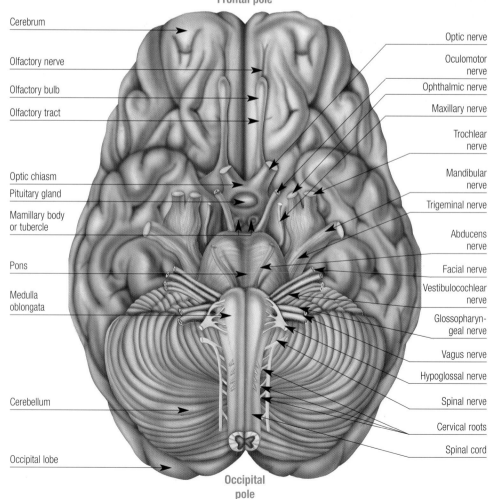

Frontal pole

Cerebrum

Olfactory nerve

Olfactory bulb

Olfactory tract

Optic chiasm

Pituitary gland

Mamillary body or tubercle

Pons

Medulla oblongata

Cerebellum

Occipital lobe

Optic nerve

Oculomotor nerve

Ophthalmic nerve

Maxillary nerve

Trochlear nerve

Mandibular nerve

Trigeminal nerve

Abducens nerve

Facial nerve

Vestibulocochlear nerve

Glossopharyn-geal nerve

Vagus nerve

Hypoglossal nerve

Spinal nerve

Cervical roots

Spinal cord

Occipital pole

## Meninges

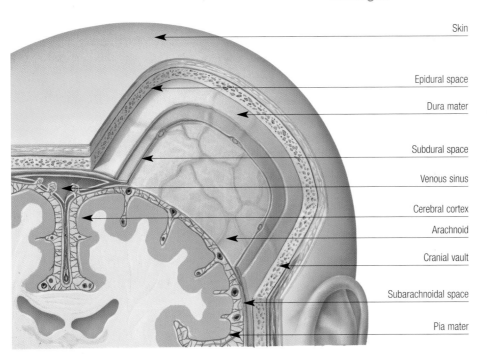

Skin

Epidural space

Dura mater

Subdural space

Venous sinus

Cerebral cortex

Arachnoid

Cranial vault

Subarachnoidal space

Pia mater

The meninges are three superimposed, concentric membranes that envelop the brain and the spinal cord and have a protective function. The dura mater, which is the outer, bulky, and strong membrane, is situated in immediate contact with the internal surface of the cranium and the internal walls of the vertebral canal, where the spinal cord is located. The arachnoid is the intermediate membrane. It is flexible and elastic. Its structure forms a mesh reminiscent of a spider web. The pia mater, the internal membrane, is very fine and delicate, firmly adhering to the surface of the brain and of the spinal cord.

## Detail of meninges

Among the various meninges, as among the dura mater and the bones that it covers on the inside, are various spaces that have different names and characteristics. The subarachnoid space, which separates the arachnoid from the pia mater, is occupied by cerebrospinal fluid. The subdural space is situated between the dura mater and the arachnoid. The epidural space is situated between the dura mater and the surface of the bones it covers on the inside. In some sectors, where the dura mater is divided into two sheets, spaces form that are filled with blood, called venous sinuses. Among these are arachnoid branches, called arachnoid granulations, responsible for filtering the cerebrospinal fluid.

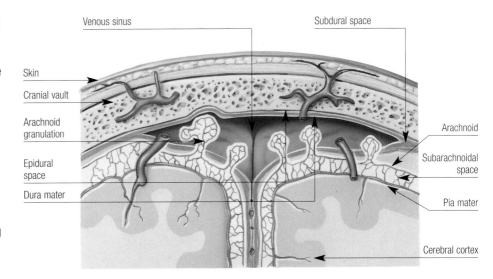

Venous sinus

Subdural space

Skin

Cranial vault

Arachnoid granulation

Epidural space

Dura mater

Arachnoid

Subarachnoidal space

Pia mater

Cerebral cortex

## Cerebral ventricles

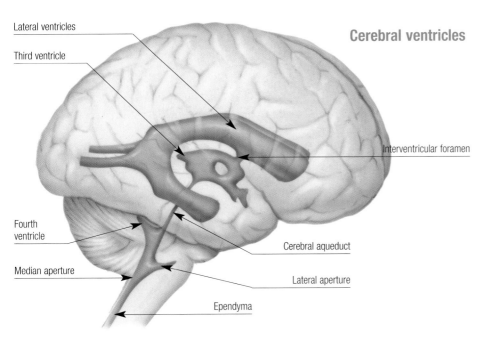

Lateral ventricles

Third ventricle

Interventricular foramen

Fourth ventricle

Median aperture

Cerebral aqueduct

Lateral aperture

Ependyma

The interior of the brain contains various cavities filled with cerebrospinal fluid linked to each other by fine canals and openings permitting the secretion of the fluid. Lateral ventricles are situated within the two cerebral hemispheres. The third ventricle is practically in the center of the cerebrum. The fourth ventricle is situated between the brain stem and the cerebellum, communicating with the former through the cerebral aqueduct and also with the subarachnoidal space, which continues downward with the central canal of the spinal cord, the ependyma.

# Cerebrum

**The cerebrum is the most important organ of the nervous** system. All voluntary activity is generated and higher mental functions, such as thinking, intelligence, memory, and language, are processed in the cerebral cortex. The thin layer of gray matter that constitutes the external surface of the organ is made up of millions of neurons. This is also where sensations are consciously perceived.

## Top view of cerebrum

The cerebrum has a very complex structure. It has millions of neurons whose cell bodies are grouped in some sectors constituting the so-called gray matter. Other sectors contain nerve fibers covered with myelin that emerge from them or extend to those coming from other areas and constitute the so-called white matter. The cerebrum is divided into two symmetrical halves, the cerebral hemispheres, separated by a large, longitudinal fissure whose outer surface is represented by a layer of gray matter with a thickness of 3 to 4 mm made up of various strata of neuronal bodies: the cerebral cortex.

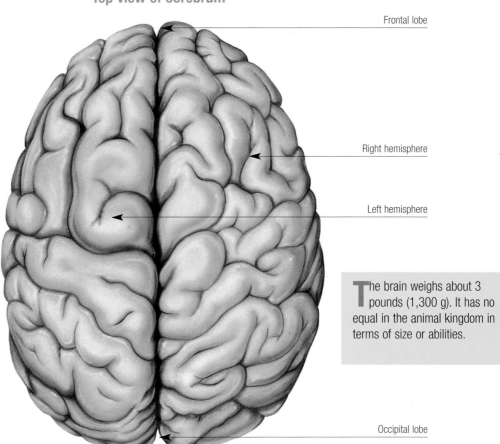

Frontal lobe

Right hemisphere

Left hemisphere

Occipital lobe

The brain weighs about 3 pounds (1,300 g). It has no equal in the animal kingdom in terms of size or abilities.

## External view of cerebrum

The surface of the cerebrum is highly irregular since the cortex features a multitude of folds that form numerous grooves and clefts. The deepest clefts are called fissures or sulci. They divide each cerebral hemisphere into four sectors called lobes, whose name corresponds to that of the cranial bone that covers them: the frontal, parietal, temporal, and occipital lobes. Each lobe is grooved by shallower clefts that constitute the boundary of some elongated areas called circumvolutions.

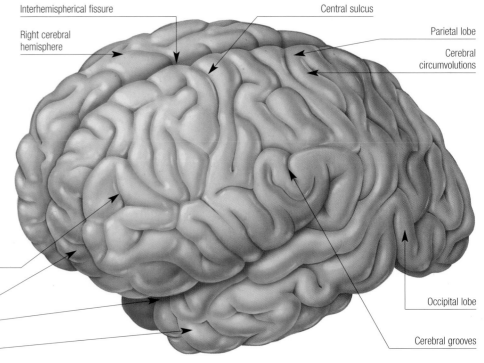

Interhemispherical fissure

Right cerebral hemisphere

Central sulcus

Parietal lobe

Cerebral circumvolutions

Left cerebral hemisphere

Frontal lobe

Lateral sulcus

Temporal lobe

Occipital lobe

Cerebral grooves

## View of internal face of cerebrum

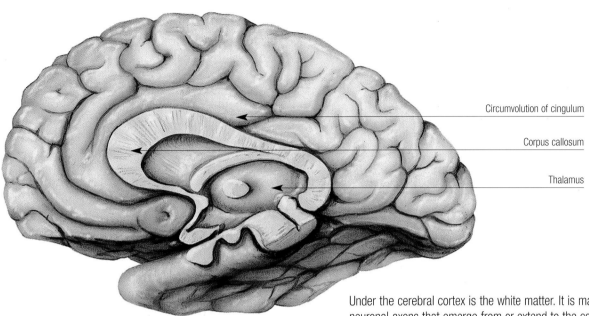

Circumvolution of cingulum

Corpus callosum

Thalamus

Under the cerebral cortex is the white matter. It is made up of the neuronal axons that emerge from or extend to the cortex. It joins different areas of one hemisphere (association fibers), areas of the cortex with different encephalic structures (projection fibers), and also the two hemispheres to each other (commissural fibers). Fibers that link both hemispheres are grouped in a thick band of white matter called the corpus callosum.

## Cross section of cerebrum

In the deepest part of the cerebrum are accumulations of neuronal bodies that constitute the gray matter. They include the following: the thalamus, the caudate nucleus, the lenticular nucleus, which is formed by the pallid nucleus and the putamen, or the hypothalamus, below which hangs the pituitary. These are separated from each other by strips of white matter where a thin layer called the external capsule stands out and where nerve fibers connect the cerebral cortex to the thalamus, the brain stem, and the spinal cord.

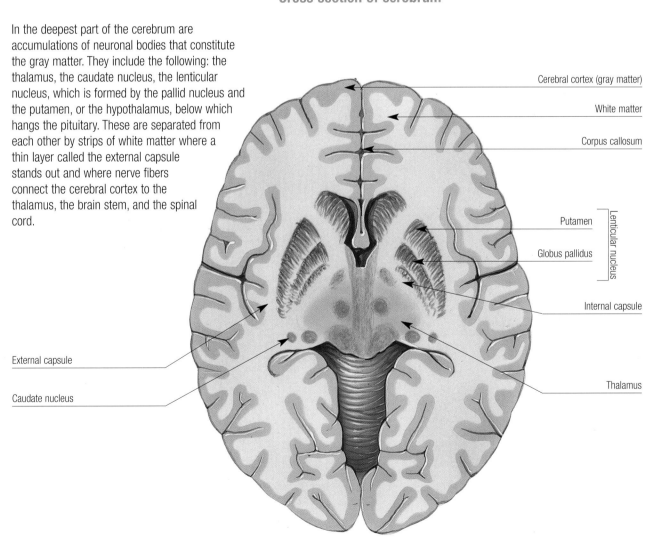

Cerebral cortex (gray matter)

White matter

Corpus callosum

Putamen

Globus pallidus

Lenticular nucleus

Internal capsule

Thalamus

External capsule

Caudate nucleus

# Cerebellum and spinal cord

**The cerebellum and the spinal cord are parts of the central** nervous system. The cerebellum has multiple connections with other sectors of the brain and modulates body movements. The spinal cord is an extension of the brain and is housed in the vertebral column. The sensory and motor nerves originate in the spinal cord.

## Cerebellum

The shape of the cerebellum is round, recalling the shape of a butterfly. It consists of two lateral parts, the cerebellar hemispheres linked by a longitudinal central part called the vermis. On its surface, it features deep parallel fissures that run from the center to the periphery of the hemispheres and divide the organ into various lobes that are grooved by numerous shallow folds. It is connected with the spinal cord and with the cerebrum through the brain stem via three bulky bundles of nerve fibers known as cerebellar peduncles. The organ, as a matter of fact, is a passage route for all the sensory and motor information concerning body balance and movements.

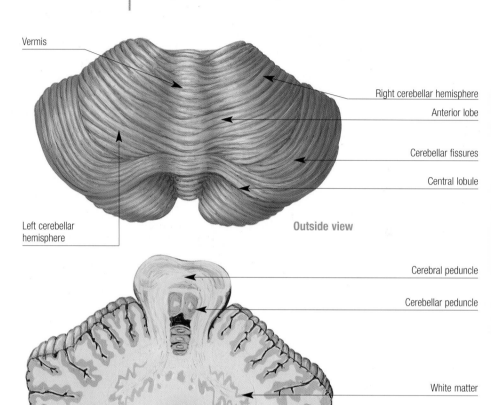

Vermis

Right cerebellar hemisphere

Anterior lobe

Cerebellar fissures

Central lobule

Left cerebellar hemisphere

**Outside view**

Cerebral peduncle

Cerebellar peduncle

White matter

Vermis

Gray matter

**Cross section**

## Spinal cord

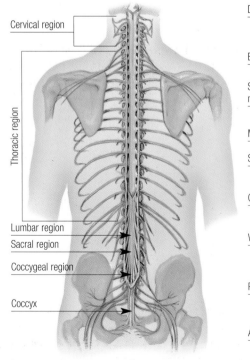

Cervical region

Thoracic region

Lumbar region

Sacral region

Coccygeal region

Coccyx

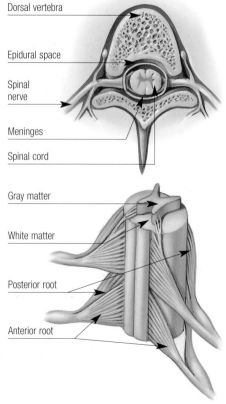

Dorsal vertebra

Epidural space

Spinal nerve

Meninges

Spinal cord

Gray matter

White matter

Posterior root

Anterior root

The spinal cord is a prolongation of the brain, a long cylinder contained in the interior of the vertebral column and from which emerge the peripheral nerves. It springs up in the medulla oblongata and runs along the interior of the central canal of the vertebral column to the lumbar region of the vertebral column. Although there is no interruption in it, it is believed that the spinal cord is divided into five regions, each of which takes the name of the part of the vertebral column through which the respective spinal nerves run: the cervical, thoracic, lumbar, sacral, and coccygeal regions. Since the length of the spinal cord is less than that of the vertebral column that contains it, the lower part of the internal canal of the vertebral column is grooved by the nerve roots that emerge in the lower sectors, an assembly that is called the coccyx or tailbone.

## Cross section of spinal cord

Posterior median sulcus

Posterior horn

White matter
(posterior sensory cords)

Gray matter

Anterior horn

White commissure

Anterior median fissure

White matter
(anterior motor cords)

Gray commissure

Ependyma

A transverse cross section shows that the central part of the spinal cord is shaped like a butterfly. This area is made up of gray matter and contains numerous nerve cell bodies. The gray matter is surrounded white matter made up of nerve fiber bundles that run the length of the spinal cord. Some of these bundles carry sensory information from the periphery to the brain. Others transport motor impulses from the brain to the periphery. These fibers are grouped in various fasciculi or cords and in a specific order. Those that carry motor information are situated ventrally. Those that carry sensory information are situated dorsally.

## Schematic illustration of the spinal cord with the spinal nerves

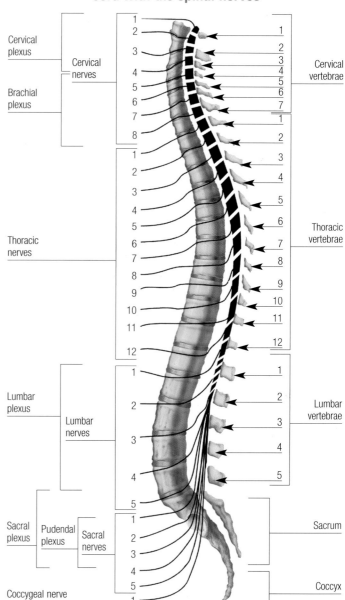

Cervical plexus

Cervical nerves

Brachial plexus

Thoracic nerves

Lumbar plexus

Lumbar nerves

Sacral plexus

Pudendal plexus

Sacral nerves

Coccygeal nerve

Cervical vertebrae

Thoracic vertebrae

Lumbar vertebrae

Sacrum

Coccyx

## Schematic illustration of the spinal cord and a spinal nerve

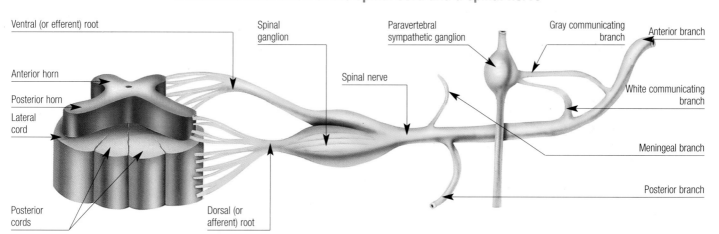

Ventral (or efferent) root

Spinal ganglion

Paravertebral sympathetic ganglion

Gray communicating branch

Anterior branch

Anterior horn

Posterior horn

Lateral cord

Spinal nerve

White communicating branch

Meningeal branch

Posterior branch

Posterior cords

Dorsal (or afferent) root

From the medulla oblongata spring 31 pairs of rachidian or spinal nerves whose branches extend to all of the sectors of the body. Each one comes from a segment of the medulla oblongata where an anterior root originates on each side. The root is made up of the axons of the motor neurons located in the anterior horn and the posterior root made up of the axons that conduct the sensory stimuli of the skin and the internal organs. Each posterior root features a thickening called the spinal or rachidian ganglion, where the sensory stimuli arrive and where the axons penetrate the medulla oblongata via the posterior part.

# Peripheral nervous system

**The peripheral nervous system is made up of the nerves that** emerge from the brain and from the spinal cord. They gather stimuli coming from the outside and from the body itself. They also carry orders from the brain to the rest of the body.

## Nerve structure

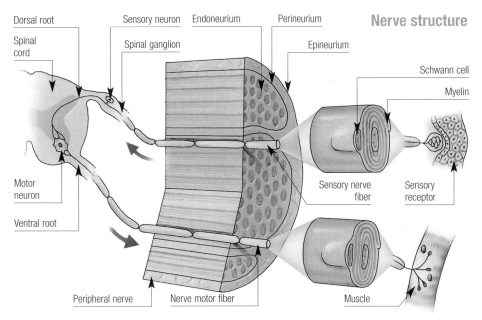

Dorsal root
Spinal cord
Sensory neuron
Spinal ganglion
Endoneurium
Perineurium
Epineurium
Schwann cell
Myelin
Sensory nerve fiber
Sensory receptor
Motor neuron
Ventral root
Peripheral nerve
Nerve motor fiber
Muscle

A nerve consists of a group of nerve fibers, in other words, axons or prolongations of neurons, plus neuroglial cells and other connective cells that protect and maintain the nerve. The nerve fibers are grouped in bundles enveloped in sheaths of connective tissue. The various bundles that constitute a nerve are surrounded by an external envelope called an epineurium.

## Diagram illustrating the spinal nerves and their branches

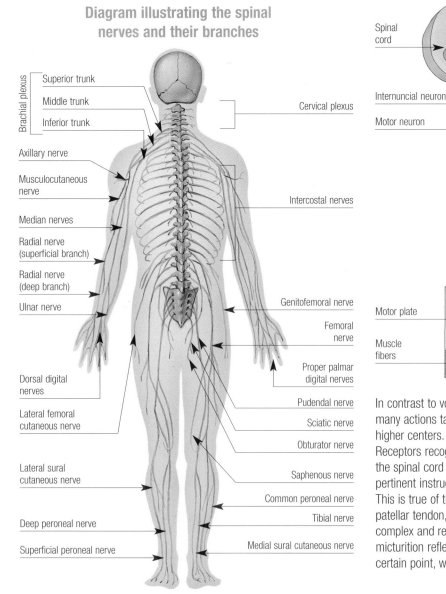

Brachial plexus
Superior trunk
Middle trunk
Inferior trunk
Axillary nerve
Musculocutaneous nerve
Median nerves
Radial nerve (superficial branch)
Radial nerve (deep branch)
Ulnar nerve
Dorsal digital nerves
Lateral femoral cutaneous nerve
Lateral sural cutaneous nerve
Deep peroneal nerve
Superficial peroneal nerve

Cervical plexus
Intercostal nerves
Genitofemoral nerve
Femoral nerve
Proper palmar digital nerves
Pudendal nerve
Sciatic nerve
Obturator nerve
Saphenous nerve
Common peroneal nerve
Tibial nerve
Medial sural cutaneous nerve

## The reflex action

Spinal cord
Sensory neuron
Spinal ganglion
Dorsal root
Internuncial neuron
Ventral root
Motor neuron
Muscle
Motor plate
Axon
Myelin
Muscle fibers

In contrast to voluntary actions that are controlled by the cerebrum, many actions take place automatically without the participation of the higher centers. They take place through a circuit called the reflex arc. Receptors recognize the stimulus, nerve fibers carry the information to the spinal cord where the response is processed, and fibers pass the pertinent instructions on to the organs charged with carrying them out. This is true of the patellar reflex. In response to percussion on the patellar tendon, the leg automatically extends. Other reflexes are more complex and require participation of the brain stem. For example, the micturition reflex is triggered when the bladder is full of urine. Up to a certain point, we can voluntarily inhibit urination.

## Cranial nerves

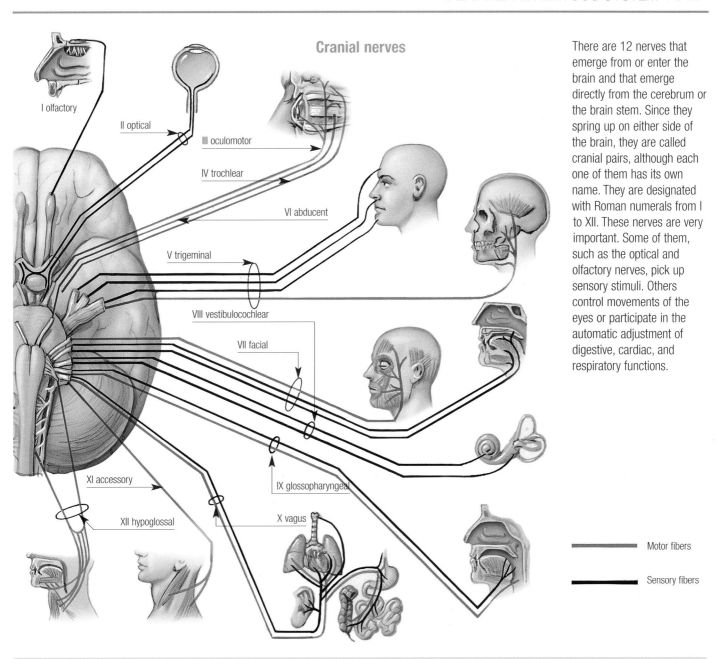

I olfactory

II optical

III oculomotor

IV trochlear

VI abducent

V trigeminal

VIII vestibulocochlear

VII facial

XI accessory

IX glossopharyngeal

XII hypoglossal

X vagus

Motor fibers

Sensory fibers

There are 12 nerves that emerge from or enter the brain and that emerge directly from the cerebrum or the brain stem. Since they spring up on either side of the brain, they are called cranial pairs, although each one of them has its own name. They are designated with Roman numerals from I to XII. These nerves are very important. Some of them, such as the optical and olfactory nerves, pick up sensory stimuli. Others control movements of the eyes or participate in the automatic adjustment of digestive, cardiac, and respiratory functions.

## Cranial nerve function

| Nerves | Name | Function |
|--------|------|----------|
| Pair I: | Olfactory | Conducts olfactory sensations from the nasal fossa to the cerebrum |
| Pair II: | Optical | Conducts visual sensations from the retina of the eye to the cerebrum |
| Pair III: | Oculomotor | Participates in the control of eye movements |
| Pair IV: | Trochlear | Participates in the control of eye movements |
| Pair V: | Trigeminal | Picks up stimuli from the face and participates in the control of mastication |
| Pair VI: | Abducent | Participates in the control of eye movements |
| Pair VII: | Facial | Controls the movements of the face muscles and passes gustatory sensations from the tongue to the cerebrum |
| Pair VIII: | Vestibulocochlear | Conducts auditory sensations and stimuli that allow control of equilibrium from the inner ear to the cerebrum |
| Pair IX: | Glossopharyngeal | Controls the movements of pharynx muscles and passes gustatory sensations from the tongue to the cerebrum |
| Pair X: | Vagus | Controls the movements of the pharynx and the larynx and participates in the regulation of the neck organs, chest (heart, respiration), and the abdomen (digestive system) |
| Pair XI: | Accessory | Controls the movements of the neck muscles, the shoulder, and the larynx |
| Pair XII: | Hypoglossal | Controls the movements of the tongue |

# Cerebral cortex and nerve paths

pathways. Orders processed in the brain travel to the peripheral nervous system. Stimuli from the outside and from the body travel to the central nervous system.

## Brain areas

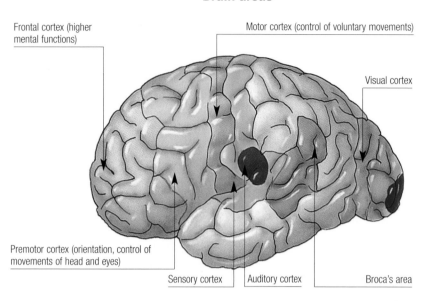

Frontal cortex (higher mental functions)

Motor cortex (control of voluntary movements)

Visual cortex

Premotor cortex (orientation, control of movements of head and eyes)

Sensory cortex

Auditory cortex

Broca's area

Although much remains to be learned about the way the cerebrum works, it has been possible to locate the areas of the cerebrum that are responsible for distinct and varied functions. Thus, we know that voluntary movements originate in the motor cortex located in the ascending frontal circumvolution, where we also find that each sector is specifically associated with the mobility of each part of the body. About the same happens with sensory information. Tactile stimuli, pain stimuli, temperature, and various stimuli coming from the body are perceived in the ascending parietal circumvolution. We also have located various sense areas, such as the visual cortex located in the occipital lobe and the auditory cortex located in the temporal area. We even have located some areas where higher intellectual functions, such as speech, are processed.

## Brain Motor and Sensory Cortices

Primary motor area

Primary somatosensory area

Motor cortex

Sensory cortex

## Lateral

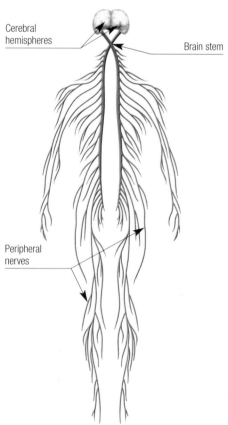

Cerebral hemispheres

Brain stem

Peripheral nerves

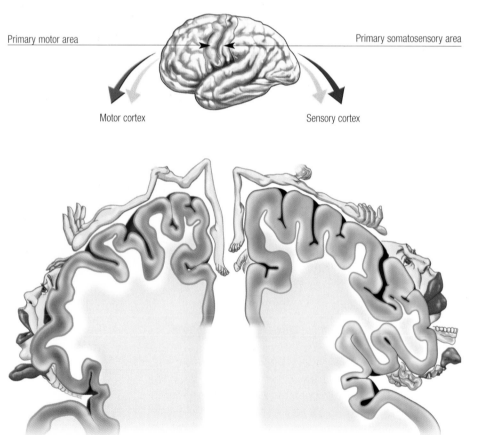

In the motor cortex located in the primary motor area and in the sensory cortex located in the primary somatosensory area on either side of the central sulcus is an extremely strong correlation between each sector and the corresponding body part. The association is so specific that a human caricature could be drawn that represents the approximate sectors within the brain that control sensory and motor activities of the corresponding body parts.

*Both the motor nerve paths and the sensory nerve paths cross each other on the way through the brain stem. Therefore, each cerebral hemisphere controls movements of and receives information from the opposite side of the body.*

## Motor pathways

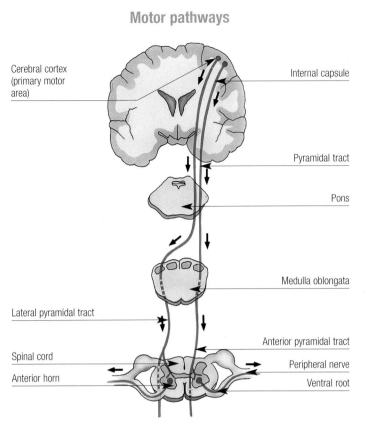

Cerebral cortex (primary motor area)

Internal capsule

Pyramidal tract

Pons

Medulla oblongata

Lateral pyramidal tract

Spinal cord

Anterior horn

Anterior pyramidal tract

Peripheral nerve

Ventral root

## Sensory pathways

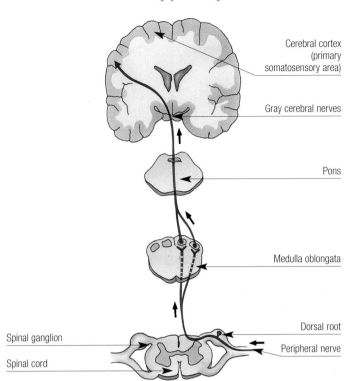

Cerebral cortex (primary somatosensory area)

Gray cerebral nerves

Pons

Medulla oblongata

Dorsal root

Spinal ganglion

Spinal cord

Peripheral nerve

The motor cortex contains an accumulation of neurons known as pyramidal cells, responsible for every motor action. Their axons form the so-called pyramidal paths that are directed to the brain stem. In the spinal cord, the nerve fibers are divided into two branches. Some cross to the other side and form the lateral pyramidal tract that descends along the lateral side of the spinal cord. The remainder constitute the anterior pyramidal tract that descends along the anterior side of the spinal cord. Here, contact is established with neurons whose axons are part of the peripheral nerves that extend all the way to the skeletal muscles.

Stimuli of various kinds coming from the outside (tactile, painful, thermal) and from the interior of the body (muscular, tendinous, articular) are recorded by special receptors that trigger nerve impulses whose destination is the central nervous system. These impulses travel along sensitive nerve fibers that extend all the way to the ganglia of the dorsal roots where the neuronal bodies are located. Then the stimuli continue on their way along these routes and penetrate the spinal cord following an ascending path via specific cords, depending on the type of information they transmit until they reach various encephalic structures. After distinct stops, they finally arrive at the primary somatosensory area of the cerebral cortex, where sensations are actually consciously perceived.

## Cerebral control of a right-handed individual

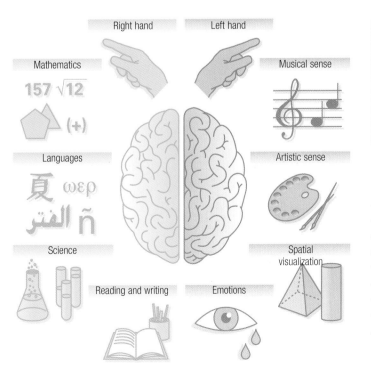

Right hand

Left hand

Mathematics

157 √12

Musical sense

Languages

夏 ωερ
الفتر ñ

Artistic sense

Science

Spatial visualization

Reading and writing

Emotions

### Pain

The pain receptors detect pain stimuli. They are found in different body tissues, such as the skin, the viscera, the blood vessels, the muscles, the capsules of the connective tissue, the periosteum, and the falx cerebri. The other tissues barely have any pain receptor endings.

*In addition to controlling motor function and information from the opposite side of the body, each cerebral hemisphere shelters areas that have specific mental functions. For example, in a right-handed individual, the right hemisphere is the seat of the musical and artistic senses, of spatial visualization, and of emotions. The left hemisphere controls language, logic, and analytical abilities.*

# Autonomic nervous system

**The autonomic, or vegetative, nervous system automatically** and unconsciously, independently of our will, and without our noticing it regulates the functioning of the internal viscera, the glands, the blood vessels, and the numerous tissues of the body to enable us to adapt to the changing necessities of each moment.

## Structure of the autonomic nervous system

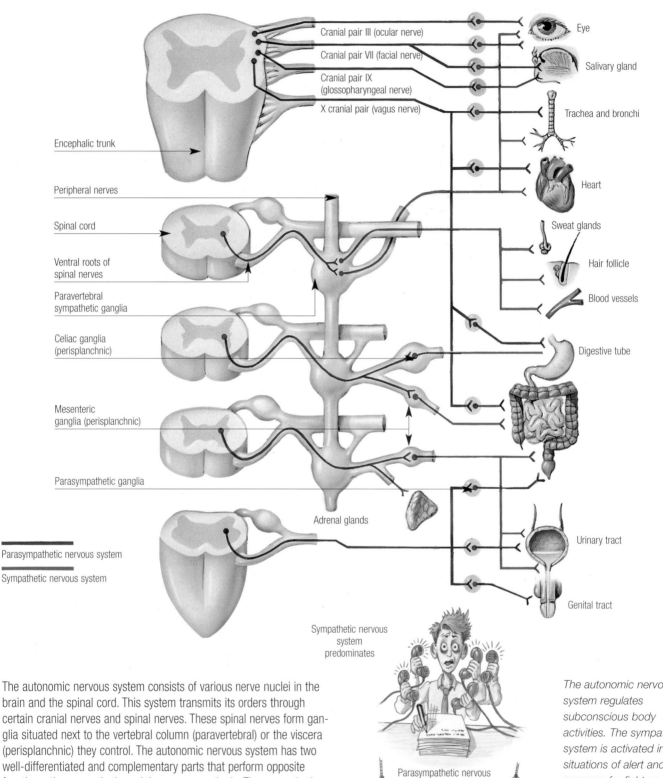

Cranial pair III (ocular nerve)

Cranial pair VII (facial nerve)

Cranial pair IX (glossopharyngeal nerve)

X cranial pair (vagus nerve)

Encephalic trunk

Peripheral nerves

Spinal cord

Ventral roots of spinal nerves

Paravertebral sympathetic ganglia

Celiac ganglia (perisplanchnic)

Mesenteric ganglia (perisplanchnic)

Parasympathetic ganglia

Adrenal glands

Parasympathetic nervous system

Sympathetic nervous system

Eye

Salivary gland

Trachea and bronchi

Heart

Sweat glands

Hair follicle

Blood vessels

Digestive tube

Urinary tract

Genital tract

Sympathetic nervous system predominates

Parasympathetic nervous system predominates

The autonomic nervous system consists of various nerve nuclei in the brain and the spinal cord. This system transmits its orders through certain cranial nerves and spinal nerves. These spinal nerves form ganglia situated next to the vertebral column (paravertebral) or the viscera (perisplanchnic) they control. The autonomic nervous system has two well-differentiated and complementary parts that perform opposite functions, the sympathetic and the parasympathetic. The sympathetic nervous system, modulated by encephalic structures, originates in the thoracic and lumbar regions of the spinal cord. The parasympathetic nervous system originates in various nuclei in the brain stem and in the sacral region of the spinal cord.

*The autonomic nervous system regulates subconscious body activities. The sympathetic system is activated in situations of alert and prepares for fight or flight. The parasympathetic system predominates at moments of relaxation and tranquility.*

## Actions of the autonomic nervous system

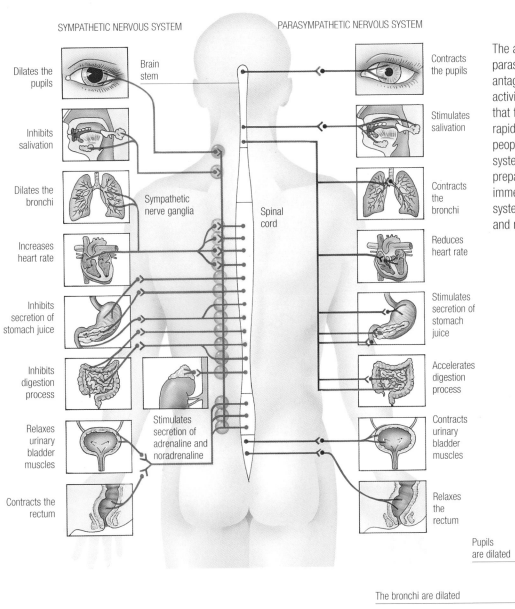

SYMPATHETIC NERVOUS SYSTEM

Dilates the pupils

Inhibits salivation

Dilates the bronchi

Increases heart rate

Inhibits secretion of stomach juice

Inhibits digestion process

Relaxes urinary bladder muscles

Contracts the rectum

Brain stem

Sympathetic nerve ganglia

Spinal cord

Stimulates secretion of adrenaline and noradrenaline

PARASYMPATHETIC NERVOUS SYSTEM

Contracts the pupils

Stimulates salivation

Contracts the bronchi

Reduces heart rate

Stimulates secretion of stomach juice

Accelerates digestion process

Contracts urinary bladder muscles

Relaxes the rectum

The actions of the sympathetic and parasympathetic nervous system are antagonistic. If one system stimulates an activity, the other one will inhibit it. Because of that fact and due to a delicate balance, they rapidly adapt the body to multiple situations people confront daily. The sympathetic nervous system is activated in situations of alert and prepares the body to confront an emergency immediately. The parasympathetic nervous system predominates in situations of relaxation and rest, such as during sleep.

### Effects of adrenaline

Pupils are dilated

The bronchi are dilated

The heart rate increases

Blood pressure goes up

Muscles tense

Vellus is erected

Bladder is contracted

The blood vessels of the muscles are dilated

### The adrenal medulla

Among its anatomical structures, the sympathetic nervous system has two very unique ones: the adrenal medullas. They are found in the center of the adrenal glands. In response to an emergency situation, each adrenal medulla goes into immediate action. They act as true sympathetic ganglia but in a very special manner. In response to signals from the neurons located in the spinal cord, they release the neurotransmitters adrenaline and noradrenaline into the blood. These neurotransmitters circulate and reach all sectors of the body, triggering the sympathetic nervous system to prepare the individual for fight or flight.

The sympathetic nervous system fibers extend all the way to the various tissues, where they exert their influence. There, in response to timely stimuli, they release two neurotransmitters, adrenaline and noradrenaline, for which there are specific receptors that trigger the proper responses. In a stressful situation—an event that causes fear or requires an immediate response, it has a special mechanism. In the face of danger, the sympathetic system produces a stimulus to the adrenal glands to release a hormone called adrenaline into the circulation. Through the blood, adrenaline reaches the entire body and gives rise to a series of adjustments that usually make it possible to react to an emergency with great efficiency.

# Kidneys and production of urine

**The urinary system filters the blood and expels the wastes. The** kidneys remove excess water, excess salts, metabolic wastes, and toxins from the blood. If the urinary system does not function properly, these items accumulate in the blood and harm the body.

## Components of urinary system

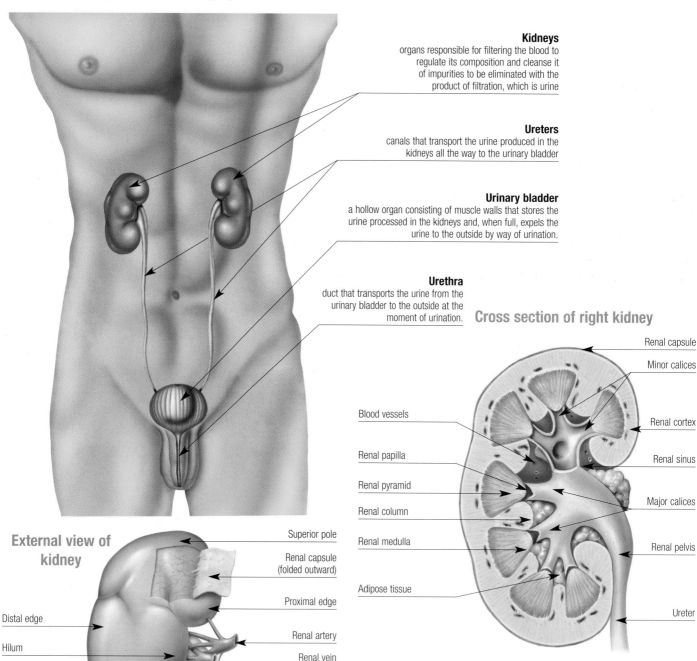

**Kidneys**
organs responsible for filtering the blood to regulate its composition and cleanse it of impurities to be eliminated with the product of filtration, which is urine

**Ureters**
canals that transport the urine produced in the kidneys all the way to the urinary bladder

**Urinary bladder**
a hollow organ consisting of muscle walls that stores the urine processed in the kidneys and, when full, expels the urine to the outside by way of urination.

**Urethra**
duct that transports the urine from the urinary bladder to the outside at the moment of urination.

### Cross section of right kidney

Renal capsule

Minor calices

Blood vessels

Renal cortex

Renal papilla

Renal sinus

Renal pyramid

Major calices

Renal column

Renal medulla

Renal pelvis

Adipose tissue

Ureter

### External view of kidney

Superior pole

Renal capsule (folded outward)

Proximal edge

Distal edge

Renal artery

Hilum

Renal vein

Renal pelvis

Ureter

Inferior pole

Interior face of right kidney

The kidneys are two symmetrical organs located in the highest and posterior part of the abdominal cavity on either side of the vertebral column in the lumbar region. It has a characteristic oval shape, recalling the shape of a kidney bean. It measures around 5 inches (12 cm) along its major axis with a width of about 2 inches (6 cm) and a thickness of about 1 inch (3 cm), weighing approximately 5 ounces (150 g).

A cross section of the kidney reveals two regions: the renal cortex and the renal medulla. The **renal cortex,** which is peripheral, has prolongations called renal columns. These intrude into the renal medulla and divide it into various sectors. The dark-red **renal medulla** consists of triangular-shaped renal pyramids. The tip of each pyramid—the renal papilla—projects toward the renal sinus. The renal papillae have numerous tiny orifices through which the urine can flow. The renal pyramids empty the urine through the renal papillae into thin tubes called minor calices. These calices then pass the urine to wider tubes called major calices. The major calices join to form a single funnel-shaped cavity, the renal pelvis, which travels through the hilum and becomes the ureter.

The **nephron** is the functional unit of the kidney since it is responsible for filtering the blood and processing the urine. It consists of a renal corpuscle where the blood is filtered and a collecting tubule where the production of urine ends. The renal corpuscle is made up of a glomerulus, a tangle of capillary vessels through which the blood circulates. It is surrounded by a double membrane in the form of a funnel called Bowman's capsule, which continues directly with the proximal convoluted tubule.

The **glomerulus** accounts for the branches of the afferent glomerular arteriole that carry blood to the renal corpuscle and that then link an efferent glomerular arteriole through which the already filtered blood is discharged. Between the two layers of Bowman's capsule is a tiny fissure, the urinary space, where the product of the glomerular filtrate is emptied. The continuation of this capsule, the convoluted tubules, is a canal made up of various segments of varying shape and thickness that are surrounded by blood vessels and that aid the glomerular filtrate in the processing of urine.

## Nephron

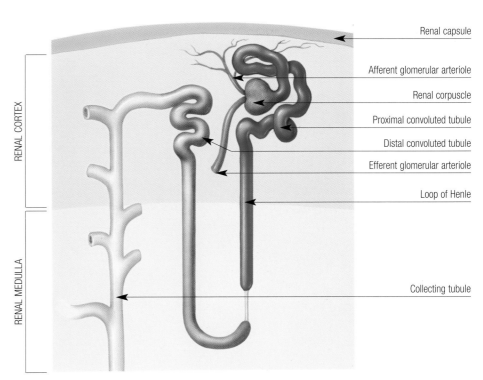

- Renal capsule
- Afferent glomerular arteriole
- Renal corpuscle
- Proximal convoluted tubule
- Distal convoluted tubule
- Efferent glomerular arteriole
- Loop of Henle
- Collecting tubule

RENAL CORTEX
RENAL MEDULLA

## Renal corpuscle

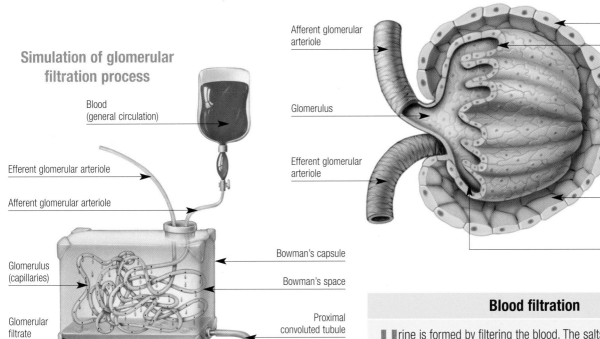

- Afferent glomerular arteriole
- Glomerulus
- Efferent glomerular arteriole
- Parietal layer
- Visceral layer
- Bowman's capsule
- Proximal tubule
- Bowman's space
- Glomerular capillaries

## Simulation of glomerular filtration process

- Blood (general circulation)
- Efferent glomerular arteriole
- Afferent glomerular arteriole
- Glomerulus (capillaries)
- Glomerular filtrate
- Bowman's capsule
- Bowman's space
- Proximal convoluted tubule

## Processing urine

The glomerular filtrate is modified as it travels through the convoluted tubules since most of the water and various substances that it contains are **reabsorbed**. In other words, these substances move into the adjacent capillary vessels and return to the blood, while others that have not been filtered in the glomerulus are **secreted** in the opposite direction. Substances pass from the blood that circulates through the nearby capillaries to the interior of the tubule. As a result, 47 gallons (180 L) of glomerular filtrate are converted into barely 1.6–2.1 quarts (1.5–2 L) of urine daily. Moreover, the body recovers **useful substances** that may have been filtered in the glomerulus to maintain an adequate physical and chemical equilibrium in the internal environment.

## Blood filtration

Urine is formed by filtering the blood. The salts, excess water, metabolic wastes, and toxins leave the blood by passing through tiny pores in the walls of the glomerular capillaries. The filtrate moves to the Bowman's space between the parietal and visceral layers of the Bowman's capsule and then continues to the convoluted tubule. Filtration is a passive process involving two antagonistic forces: hydrostatic pressure and coloidosmotic pressure. Hydrostatic pressure is the actual pressure exerted by the plasma on the glomerular capillaries. Coloidosmotic pressure, in contrast, is the power of the plasma to attract and retain proteins, blood cells, and other large substances. Because of their size, these substances cannot pass through the glomerular capillaries. As a result, they always remain in the blood.

# Urinary tract and urination

**Urine in the kidneys is transported through the ureter to the** bladder, a hollow organ with distensible walls where the fluid is accumulated and remains stored until it is eliminated to the outside through the urethra by urination.

### Urinary tract

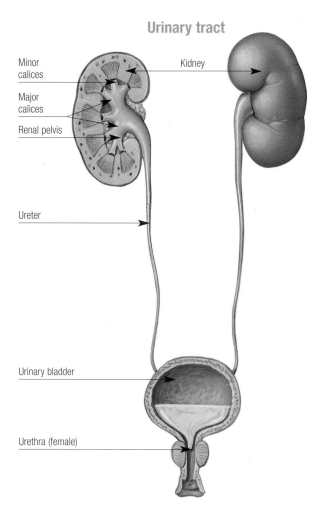

Minor calices
Major calices
Renal pelvis
Kidney
Ureter
Urinary bladder
Urethra (female)

The urinary tract is a series of interconnected hollow structures responsible for collecting the continuous secretion of urine, storing it temporarily, and emptying it to the outside of the body through urination several times a day. The urinary tract starts in the kidneys. From each kidney emerges a funnel-shaped renal pelvis that continues to a ureter. Each of the two ureters is a long, tubular canal that descends through the abdominal cavity and empties its contents into the urinary bladder found in the pelvic region. The bladder is a dilatable, hollow organ with powerful muscular walls. It stores urine until sufficiently full. The urine then evacuates the bladder and the body by flowing through the urethra.

## The bladder
This is a hollow and dilatable organ. When it is empty, it has a more or less triangular shape. As it gradually fills, it takes on an ovoid or spherical shape. Under normal conditions, in an adult, it can store as much as 12 fluid ounces (350 mL) of urine.

The bladder consists of three differentiated parts. The **cupula** is the upper part, which is lined on the outside by the peritoneum. The **body** constitutes the major part of the organ and on its posterior side shelters the two ureteric orifices where the urine coming from the kidneys arrives. The **base** is supported on the bottom of the pelvis and has the shape of a kind of funnel, the bladder neck, which empties into a single urethral orifice that links the bladder to the urethra.

### Male urethra

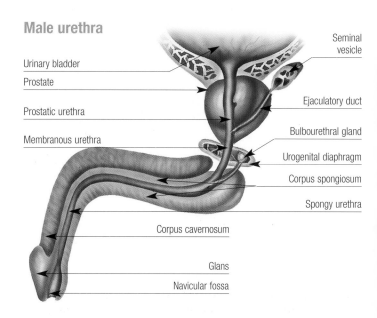

Urinary bladder
Prostate
Prostatic urethra
Membranous urethra
Seminal vesicle
Ejaculatory duct
Bulbourethral gland
Urogenital diaphragm
Corpus spongiosum
Spongy urethra
Corpus cavernosum
Glans
Navicular fossa

## Urethra
The last branch of the urinary tract is the canal through which the urine is evacuated from the bladder to outside the body. In women, the urethra has only this function. In men, it also conveys semen produced by the internal sex organs at the moment of ejaculation to the outside. The urethra begins in the urethral orifice of the bladder and ends in the external urethral orifice on the body surface.

The female urethra is 1.5–2 in. (4–5 cm) long and follows a straight, descending line, emerging on the outside in the vestibule of the vulva. The male urethra is 6–8 in. (15–20 cm) long and differentiates into three segments. The first, called the prostatic urethra, passes through the prostate. The second, the membranous urethra, runs from the prostate all the way to the root of the penis. The last one, the spongy urethra, runs through the interior of the penis inside the corpus spongiosum until it empties into the external urethral orifice at the tip of the glans.

### Section of the female urinary bladder and urethra

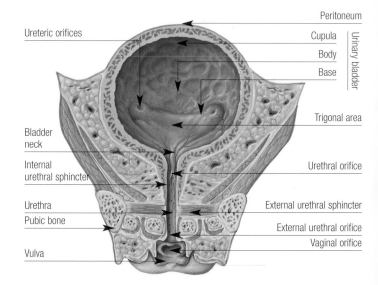

Ureteric orifices
Bladder neck
Internal urethral sphincter
Urethra
Pubic bone
Vulva
Peritoneum
Cupula
Body
Base
Urinary bladder
Trigonal area
Urethral orifice
External urethral sphincter
External urethral orifice
Vaginal orifice

## Urination mechanism

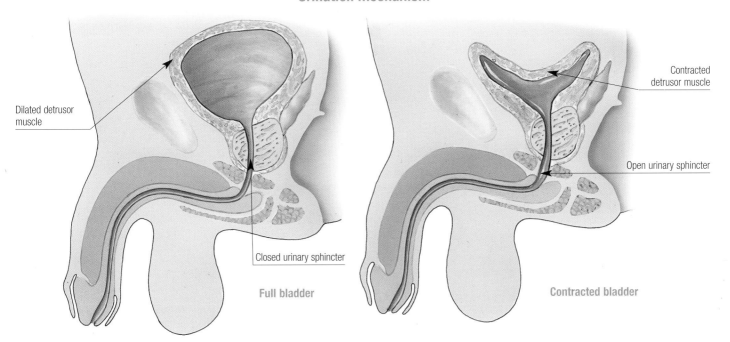

Dilated detrusor muscle

Closed urinary sphincter

**Full bladder**

Contracted detrusor muscle

Open urinary sphincter

**Contracted bladder**

The bladder stores the urine temporarily since, although its muscular walls are distensible, its fluid storage capacity is limited. When the limit is reached, urine is expelled to the outside through the urethra as a result of urination. This mechanism depends on a kind of muscular valve that is situated at the outlet of the bladder. This valve both closes the urethra so that urine cannot get to the outside and opens the urethra to allow urine to be expelled from the body.

As a matter of fact, this valve, known as the urinary sphincter, is made up of two structures that constitute barriers to the passage of the urine. The internal urethral sphincter is situated at the outlet of the bladder into the urethra, and the external urethral sphincter is located in the middle segment of the urethra. The former works automatically, but the latter can be controlled voluntarily to a certain extent. This is why it is possible to "hold" the urine until it can be voided under adequate conditions.

## Micturition reflex

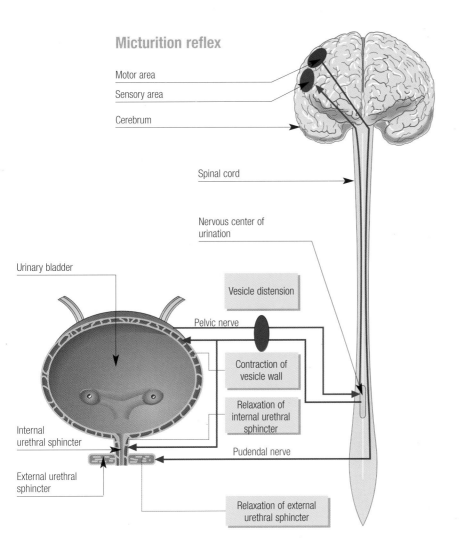

Motor area

Sensory area

Cerebrum

Spinal cord

Nervous center of urination

Urinary bladder

Vesicle distension

Pelvic nerve

Contraction of vesicle wall

Relaxation of internal urethral sphincter

Internal urethral sphincter

Pudendal nerve

External urethral sphincter

Relaxation of external urethral sphincter

Voluntary control of the external urethral sphincter is learned during the first years of life. Children learn around the age of two to identify the signals indicating that the bladder is full and how to inhibit the automatic reflex.

The bladder is emptied as a result of an automatic reflex that is triggered when the walls of the bladder distend beyond a certain limit. When that happens, nerve receptors, located in the walls of the bladder, emit a signal that goes to the urination center situated in the spinal cord, which responds with motor impulses that reach the muscular layer of the vesicle walls. At that moment, the detrusor muscle, which is a part of the bladder wall, is contracted. At the same time, the internal urethral sphincter is opened, allowing the urine to pass on to the urethra. To make sure that urine is evacuated to the outside, however, it is also necessary to relax the external urethral sphincter, which is under voluntary control.

# Male genital organs

organs and tissues. The external ones can be clearly observed. The internal ones, although in communication with the outside, are especially adapted to enable the male to carry out sexual activity and to reproduce.

## Male genital organs

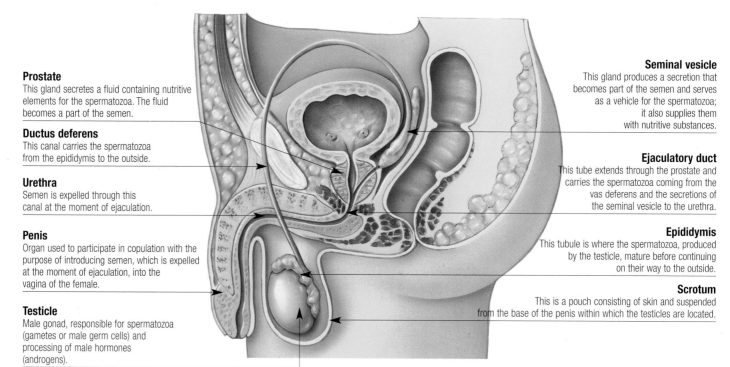

**Prostate**
This gland secretes a fluid containing nutritive elements for the spermatozoa. The fluid becomes a part of the semen.

**Ductus deferens**
This canal carries the spermatozoa from the epididymis to the outside.

**Urethra**
Semen is expelled through this canal at the moment of ejaculation.

**Penis**
Organ used to participate in copulation with the purpose of introducing semen, which is expelled at the moment of ejaculation, into the vagina of the female.

**Testicle**
Male gonad, responsible for spermatozoa (gametes or male germ cells) and processing of male hormones (androgens).

**Seminal vesicle**
This gland produces a secretion that becomes part of the semen and serves as a vehicle for the spermatozoa; it also supplies them with nutritive substances.

**Ejaculatory duct**
This tube extends through the prostate and carries the spermatozoa coming from the vas deferens and the secretions of the seminal vesicle to the urethra.

**Epididymis**
This tubule is where the spermatozoa, produced by the testicle, mature before continuing on their way to the outside.

**Scrotum**
This is a pouch consisting of skin and suspended from the base of the penis within which the testicles are located.

## External male genital organs

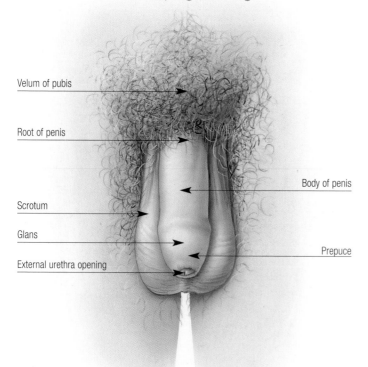

Velum of pubis

Root of penis

Scrotum

Glans

External urethra opening

Body of penis

Prepuce

## Location of male genital organs

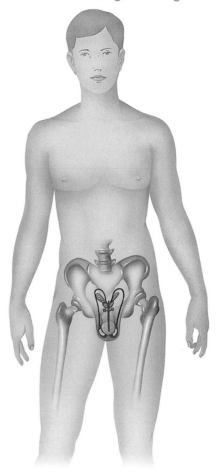

The most important male external genital organ is the **penis**, an organ through whose interior the urethra runs and that is used to participate in copulation with the purpose of introducing semen into the vagina of the female. The **scrotum** is also external. It is a cutaneous pouch that contains the testicles and holds them in the lower part of the trunk.

## Longitudinal cross section of penis

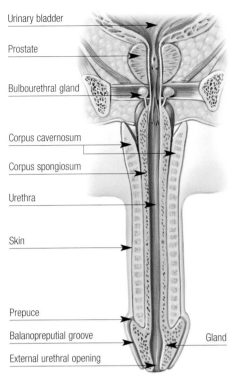

Urinary bladder

Prostate

Bulbourethral gland

Corpus cavernosum

Corpus spongiosum

Urethra

Skin

Prepuce

Balanopreputial groove

External urethral opening

Gland

The **penis** is a cylindrical organ situated in the lower part of the trunk through whose interior runs the urethra. It is divided into three parts. The root is where it is attached to the trunk. The body is the central part. A rounded end is called the glans at whose tip we find the opening of the urethra. The outside is covered by skin although with certain peculiar features. Both the root and the body of the penis are covered by skin, but the surface of the glans is covered by a delicate layer of mucosa that is very sensitive. From the boundary between the body of the penis and the glans springs the hood made up of skin called the prepuce, which covers the entire end of the penis and can be retracted.

## Cross section of penis

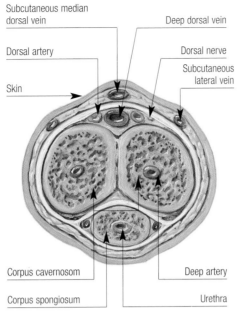

Subcutaneous median dorsal vein

Deep dorsal vein

Dorsal artery

Dorsal nerve

Subcutaneous lateral vein

Skin

Corpus cavernosom

Deep artery

Corpus spongiosum

Urethra

The unusual feature of the skin is its special capacity to change its **dimensions** and **consistency** when it enters into a state of erection. This depends on three cylindrical bodies housed inside the penis. They are made up of a special erectile tissue containing numerous connective and muscular fiber partitions that separate a large number of tiny interconnected cavities. Under certain conditions, for example, in response to a sexual stimulus, they fill with blood. Subsequently, the penis increases in size and diameter. Two of these structures are the symmetrical corpus cavernosums, situated on either side of the upper part of the body of the penis. The third is the corpus spongiosum, situated in a central position below the former and crossed longitudinally by the urethra, whose end expands in such a way that it occupies the entire interior of the gland.

## Structure of scrotum

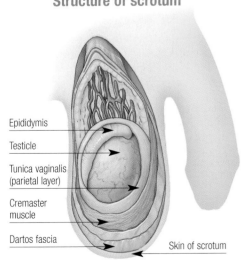

Epididymis

Testicle

Tunica vaginalis (parietal layer)

Cremaster muscle

Dartos fascia

Skin of scrotum

The **scrotum** is a pouch that hangs under the root of the penis and is where the testicles are located. This location is more suitable for the testicles than inside the abdominal cavity since the temperature to which they are exposed there is slightly lower than the temperature inside the body, and is appropriate for the production of spermatozoa. To be able to accomplish this important mission, the wall of the scrotum is made up of various layers. The outermost layer is the skin. It is fine and wrinkled with more or less deep grooves. Underneath is a muscular layer whose degree of contraction or relaxation changes the depth of the skin grooves and serves to adjust the temperature to which the testicles are exposed.

## Secondary male sex characteristics

These begin developing at puberty due to hormonal influence. Soft body hair is more abundant in males than in females and also grows on the chest (1). The pubic hair has a rhomboid shape and tends to spread almost to the navel (2). The hairline varies in width, especially in the lateral areas (3). The beard and the mustache appear on the face (4). When looking at the body in general, the male's musculature tends to develop more than the female's, and the shoulders and the back are wider (5). On the other hand, the male's hips are narrower (6). Looking at both sexes, we note a distinct distribution of subcutaneous fat deposits, which in the male tends to accumulate in the abdomen (7) and not in the hips and the muscles, as happens in the case of the female. Finally, the influence of the male sex hormones causes a development of the larynx (8) that makes the tone of the male's voice lower than that of the female.

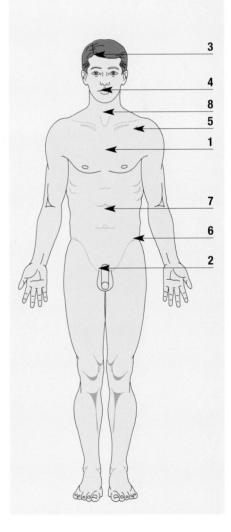

# Male genital organs

**The internal male genital organs are responsible for producing** male reproductive cells, the spermatozoa, as well as the seminal fluid, which serves as a vehicle and supplies nutrition for the secretion in which they float, called semen. The semen transports the sperm to the urethra so that it can exit to the outside at the moment of ejaculation.

## Testicle and epididymis

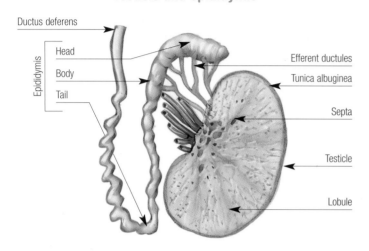

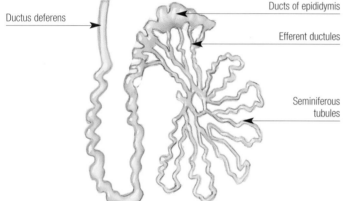

## Testicles

These are two ovoid, glandular organs with a major axis measuring 1.5–2 inches (4–5 cm) in length in the adult, located symmetrically in the lower part of the trunk inside the scrotum. Each testicle is surrounded by a fibrous membrane called the tunica albuginea. Its interior is divided into several lobes that are separated by connective tissues that enclose a number of seminiferous tubules. These are very thin, serpentine, hollow canals where the spermatozoa are generated and that flow together to form a dense network from which wider canals spring, the efferent ducts that empty into the epididymis.

## Epididymis

These are small, elongated tubular organs, each located on the upper and posterior part of the corresponding testicle, where the spermatozoa mature as they continue on their way to the outside. Each epididymis is divided into three parts: the head, the body, and the tail. The head, situated at the upper end of the testicle, consists of a series of canals that are the continuation of the efferent ducts that link up and form a single conduit that is folded back upon itself and passes through the body of the epididymis and comes out through the tail, situated at the lower end of the testicle, to empty into the corresponding ductus deferens.

## Spermatogenesis

Production of spermatozoa, or **spermatogenesis**, begins in puberty and takes place in tiny seminiferous tubules. In the embryonic stage, there are numerous spermatogonia, immature male germinal cells endowed with 46 chromosomes. Due to stimulation from follicle-stimulating hormone (FSH), these cells reproduce and are transformed into primary spermatocytes, which, in turn, divide and become secondary spermatocytes. The latter, also endowed with 46 chromosomes, are divided by a special mechanism called meiosis, which gives rise to spermatids, equipped with only 23 chromosomes, half with one X sex chromosome and the other half with one Y sex chromosome. The final stage of the process takes place in the epididymis where spermatids are transformed into spermatozoa.

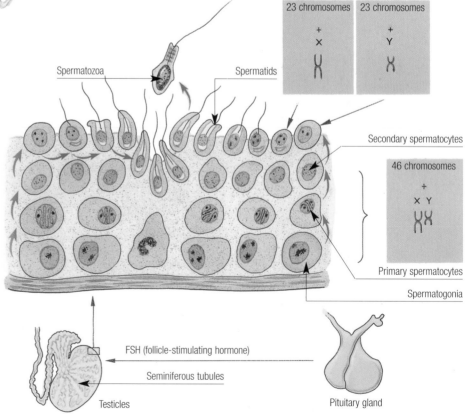

## Seminal vesicles

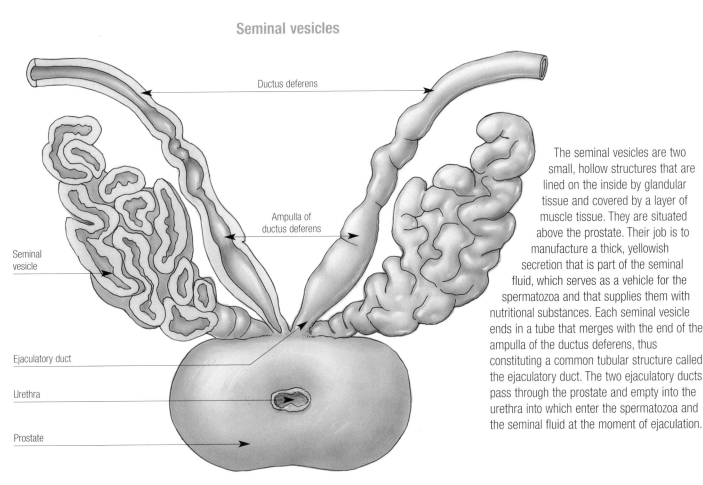

Ductus deferens

Ampulla of
ductus deferens

Seminal
vesicle

Ejaculatory duct

Urethra

Prostate

The seminal vesicles are two small, hollow structures that are lined on the inside by glandular tissue and covered by a layer of muscle tissue. They are situated above the prostate. Their job is to manufacture a thick, yellowish secretion that is part of the seminal fluid, which serves as a vehicle for the spermatozoa and that supplies them with nutritional substances. Each seminal vesicle ends in a tube that merges with the end of the ampulla of the ductus deferens, thus constituting a common tubular structure called the ejaculatory duct. The two ejaculatory ducts pass through the prostate and empty into the urethra into which enter the spermatozoa and the seminal fluid at the moment of ejaculation.

## Prostate

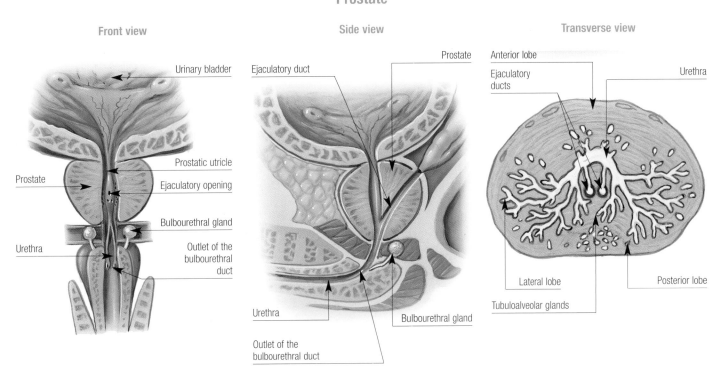

**Front view**

Urinary bladder

Prostatic utricle

Ejaculatory opening

Bulbourethral gland

Outlet of the
bulbourethral
duct

Prostate

Urethra

**Side view**

Ejaculatory duct

Prostate

Urethra

Outlet of the
bulbourethral duct

Bulbourethral gland

**Transverse view**

Anterior lobe

Ejaculatory
ducts

Urethra

Lateral lobe

Tubuloalveolar glands

Posterior lobe

The prostate is a solid organ that is situated below the urinary bladder and in front of the rectum with the first portion of the urethra passing through its center in its posterior part. The ejaculatory ducts pass through it and head for the urethra. This is essentially a glandular organ, made up of a multitude of hollow tubular structures lined on the inside by cells responsible for producing a secretion that constitutes the semen containing nutritional elements for the spermatozoa. The various tubes flow together and form a score of ducts that empty through many openings into the urethra. They deliver the prostatic secretion together with the fluid coming from the seminal vesicles and the spermatozoa coming from the testicles at the moment of ejaculation.

# Female genital organs

**The female reproductive system consists of a group of genital** organs and tissues. Some are external and can be observed directly. Others are internal, although they are in communication with the outside. They are especially adapted to allow the female to carry out sexual activity and to reproduce.

## Female genitals

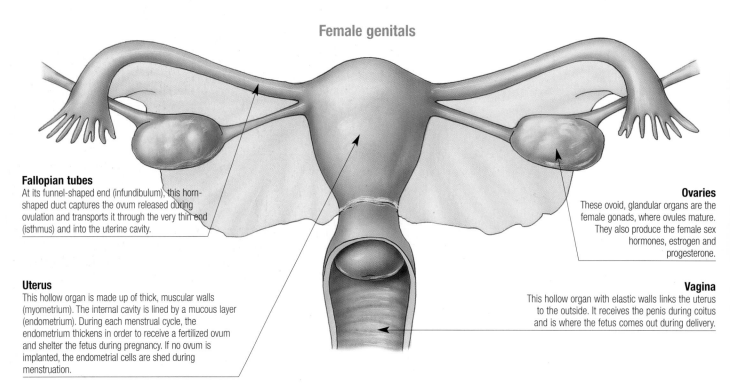

**Fallopian tubes**
At its funnel-shaped end (infundibulum), this horn-shaped duct captures the ovum released during ovulation and transports it through the very thin end (isthmus) and into the uterine cavity.

**Ovaries**
These ovoid, glandular organs are the female gonads, where ovules mature. They also produce the female sex hormones, estrogen and progesterone.

**Uterus**
This hollow organ is made up of thick, muscular walls (myometrium). The internal cavity is lined by a mucous layer (endometrium). During each menstrual cycle, the endometrium thickens in order to receive a fertilized ovum and shelter the fetus during pregnancy. If no ovum is implanted, the endometrial cells are shed during menstruation.

**Vagina**
This hollow organ with elastic walls links the uterus to the outside. It receives the penis during coitus and is where the fetus comes out during delivery.

## Vulva

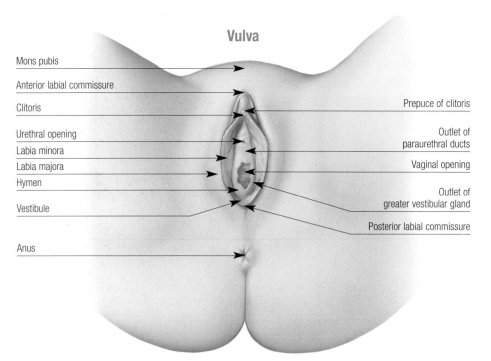

Mons pubis

Anterior labial commissure

Clitoris

Urethral opening

Labia minora

Labia majora

Hymen

Vestibule

Anus

Prepuce of clitoris

Outlet of paraurethral ducts

Vaginal opening

Outlet of greater vestibular gland

Posterior labial commissure

## Location of female genital organs

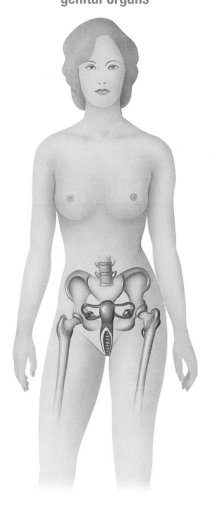

The group of external female reproductive organs located externally that are plainly visible is called the vulva. As a matter of fact, only by holding the muscles together can one see the mons pubis, a small cushion of fat situated behind the pubis that forms a prominence covered with fine hair. Farther down are the labia majora, two thick folds of skin whose external surface is also covered by fine hair. The internal surfaces of the labia majora are moistened by viscous secretions from various glands. After separating the labia majora, the labia minora can be observed. These two thinner folds of skin, devoid of any fine hair, link anteriorly and constitute the covering of the clitoris, a small erectile organ equivalent to the male penis that contains numerous sensitive nerve endings.

## Longitudinal cross section through vagina

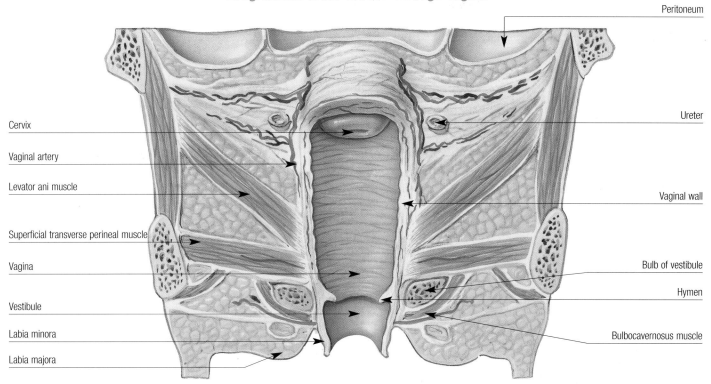

Cervix

Vaginal artery

Levator ani muscle

Superficial transverse perineal muscle

Vagina

Vestibule

Labia minora

Labia majora

Peritoneum

Ureter

Vaginal wall

Bulb of vestibule

Hymen

Bulbocavernosus muscle

The vagina is a hollow organ situated between the urinary bladder and the rectum. Its upper part communicates with the uterus, whose cervix constitutes a prominence in the bottom of the vaginal duct. The inferior end opens to the outside through an orifice situated in the vestibule of the vulva between the labia minora. In adult females, it is about 3–5 inches (8–12 cm) long and has a widely variable diameter. Its walls are very elastic and can dilate both to receive the penis during coitus and to permit the departure of the fetus at the moment of delivery. In virgin females, the orifice of the vagina is partly covered by a membrane called the hymen.

## Vaginal exam

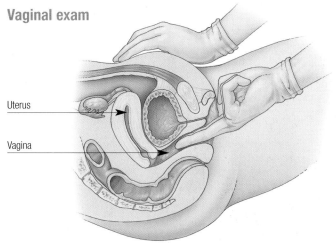

Uterus

Vagina

### Vaginal exams

The vaginal exam is the main part of a **gynecological examination**. The doctor inserts the index and middle fingers into the interior of the patient's vagina. The physician then slowly palpates the walls of the vagina and the surface of the cervix. Then the physician performs a combined vaginoabdominal touch motion. The doctor supports the free hand on the abdomen of the patient and makes gentle compressions to displace the internal genital organs upward and to palpate the uterus between both hands, determining its location, size, and consistency.

### Secondary female sex characteristics

The secondary female sex characteristics that develop in puberty vary widely. Among the most evident are the development of breasts (1), but there are also other typical characteristics. The female has softer skin and more abundant and silken hair than the male, featuring a circular hairline at the upper edge of the forehead (2), absence of small facial hair (3), much less body hair, and pubic hair that takes on a triangular shape with a straight line at the upper edge of the pubis (4). As for the development of bones and muscles, women on average are shorter than men, female musculature is less developed, and the skeleton is lighter with narrower shoulders and back (5) and wider hips (6). We also observe a distinct distribution of subcutaneous fat deposits that, in the female, tend to round the shapes of the body and to accumulate in the hips and in the thighs (7) as well as in the breasts. Finally, the female's tone of voice is higher than the male's (8).

# Female genital organs

**The internal female genital organs include the ovaries, where the** ovules are formed and where the female sex hormones are produced. The fallopian tubes are two tubular structures where fertilization occurs. The uterus is where a new human being develops after fertilization.

## Saggital cross section of the female genital system

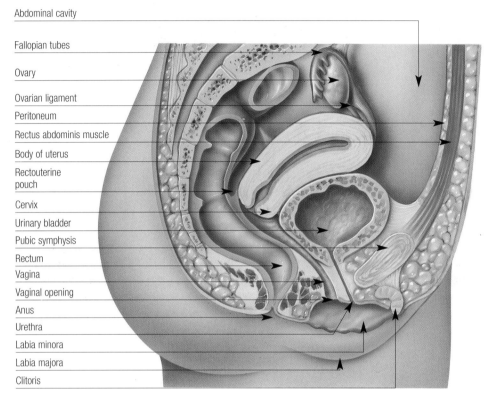

Abdominal cavity
Fallopian tubes
Ovary
Ovarian ligament
Peritoneum
Rectus abdominis muscle
Body of uterus
Rectouterine pouch
Cervix
Urinary bladder
Pubic symphysis
Rectum
Vagina
Vaginal opening
Anus
Urethra
Labia minora
Labia majora
Clitoris

## The uterus

This is a hollow organ in the shape of an inverted pear with a length of about 2.8–3 in. (7–8 cm) and a diameter of 1.2–1.6 in. (3–4 cm) at its narrowest part. It weighs something less than 3.5 oz. (100 g). Its dimensions change extraordinarily during pregnancy. The uterine body accounts for the upper part of this organ. Inside is a cavity where a fetus will develop. The cervix is the lower portion of this organ. It is in communication with the vagina. The walls of the uterus consist of three layers. The **endometrium** is a mucous layer that lines the inside of the uterine cavity. The **myometrium** is a thick layer of muscle tissue that constitutes the major part of the organ's thickness. The **perimetrium** is the thin, outer layer of connective tissue that, in most of the organ, is covered by the peritoneum.

## The fallopian tubes

These are two hollow, tubular structures shaped like a hunting horn with a length of some 4–4.7 inches (10–12 cm) and a diameter of a few millimeters. At its narrower end, each tube opens directly into the uterine cavity. The other end, the wider one, opens to the peritoneal cavity in the vicinity of the ovary. Its mission is to capture the ovum that is separated from the ovary during ovulation and to transport it to the uterus with the possibility that it might be fertilized by a spermatozoon that moves in the opposite direction. The part that is closest to the uterus is called the isthmus. The next one, becoming progressively wider, is called the ampulla. The last portion is the infundibulum, which has the shape of a funnel and whose borders are irregular and feature prolongations called fimbriae that reach toward to the ovary.

## Inside frontal view of female internal genitals

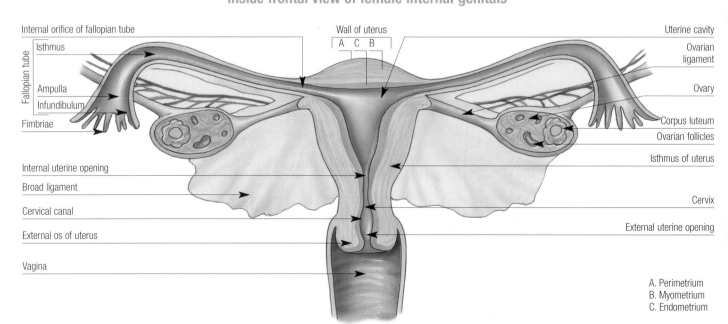

Internal orifice of fallopian tube
Fallopian tube
Isthmus
Ampulla
Infundibulum
Fimbriae
Internal uterine opening
Broad ligament
Cervical canal
External os of uterus
Vagina

Wall of uterus
A C B

Uterine cavity
Ovarian ligament
Ovary
Corpus luteum
Ovarian follicles
Isthmus of uterus
Cervix
External uterine opening

A. Perimetrium
B. Myometrium
C. Endometrium

## Internal structure of ovary

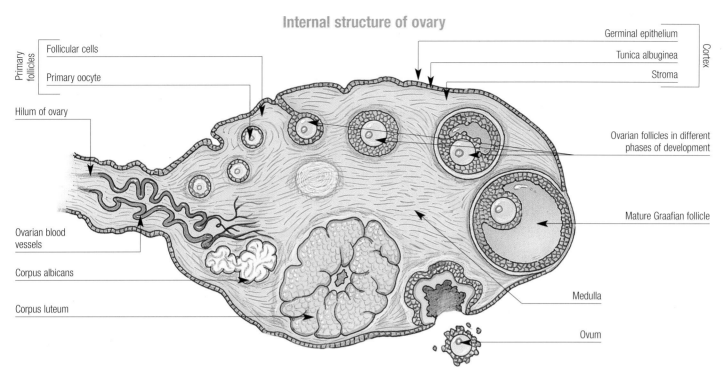

Primary follicles
- Follicular cells
- Primary oocyte

Hilum of ovary

Ovarian blood vessels

Corpus albicans

Corpus luteum

Germinal epithelium
Tunica albuginea
Stroma

Cortex

Ovarian follicles in different phases of development

Mature Graafian follicle

Medulla

Ovum

The ovaries are two glandular organs shaped like almonds with a length of 1.1–1.6 inches (3–4 cm) and a width of about 0.8 inches (2 cm) in the adult female. Each ovary has a peripheral area, the cortex, containing reproductive cells. Each also has a central area, the medulla, made up of connective tissue. The outer surface of the cortex is covered by a cellular layer called the germinal epithelium under which there is a thin, whitish, fibrous membrane called the tunica albuginea.

Below this layer we find the stroma, made up of abundant cells and connective fibers. In its thick part it contains the principal structures of the organ: the ovarian follicles, which differ in size, content, and name, depending on the degree of maturation. At the moment of birth, the ovaries contain 400,000 primary follicles. These are small in size and each is made up of a primary oocyte that constitutes the immature female germ cell surrounded by a layer of follicular cells.

## Cervicovaginal cytology

This diagnostic test, also known as the cervical smear or Papanicolau smear (Pap smear), involves taking a sample of desquamated cells, both from the mucosa that lines the vagina and from the cervix, and studying the sample microscopically. The chief usefulness of this procedure is it helps detect the presence of atypical cells coming from a precancerous lesion or from an incipient malignant tumor when there is still no other manifestation, so that any suspicion can be easily ruled out. This is a first-line preventive action in the fight against cervical cancer. This is a very simple test and entails only minor annoyance. To perform it, the physician places a speculum in the vagina through which he or she introduces a spatula, to scrape gently the walls in the back of the vagina and the surface of the cervix. The material thus obtained is placed onto a slide, fixed with solution, and sent to the laboratory and studied under the microscope. In this way, it can be determined whether the cells collected in the sample are normal or indicate some atypical risk.

### Obtaining the cervical smear sample

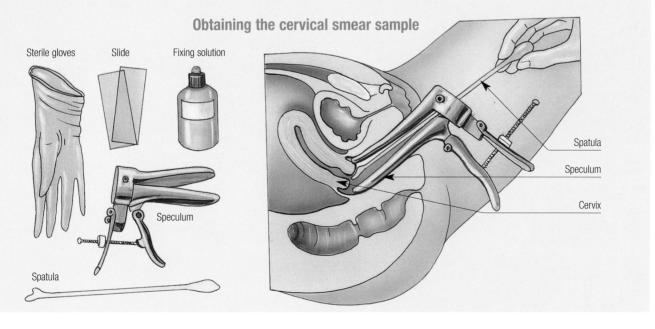

Sterile gloves    Slide    Fixing solution

Speculum

Spatula

Spatula

Speculum

Cervix

# The menstrual cycle

**The female reproductive system exerts a cyclic activity with a** duration of approximately 28 days, characterized by the regular appearance of menstrual bleeding that is repeated continuously from puberty until menopause, except during pregnancy.

## Hormonal regulation of the menstrual cycle

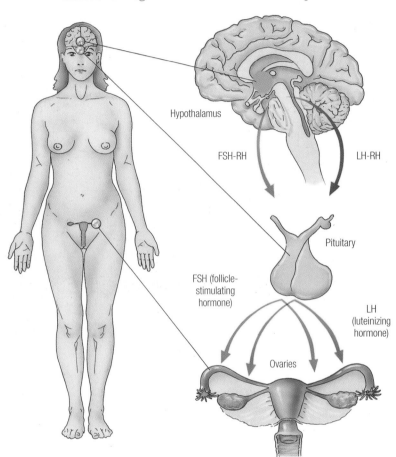

Hypothalamus

FSH-RH

LH-RH

Pituitary

FSH (follicle-stimulating hormone)

LH (luteinizing hormone)

Ovaries

The **activity** of the female reproductive system is under the control of the hypothalamus. The pituitary produces hormones that affect the ovaries and regulate the way they function. Beginning with puberty, the hypothalamus cyclically begins to produce hormones that act on the pituitary and stimulate that gland to produce hormones that modulate the way the ovaries function. These hormones are called gonadotropins: the follicle-stimulating hormone, or FSH, and the luteinizing hormone, or LH. These hormones regulate the activity of the ovaries since they stimulate the growth of ovarian follicles, the processing of female sex hormones (estrogen and progesterone), and ovulation. In turn, the hormones produced by the ovaries prepare the uterus in a cyclical fashion so that in the event of fertilization, it can receive the embryo and make the development of a pregnancy possible.

The ovaries **produce the ova** and have an endocrine function. Both of these functions are regulated by the pituitary gonadotropins. The production of ova starts in puberty when cyclically and under the influence of the FSH hormone, some of the primary follicles present in the ovaries from birth onward begin to mature. The same is true of the oocytes or immature germ cells that they contain inside. As they mature, the follicles produce estrogen that prepares the uterus for the possible reception of a fertilized ovule.

In general, only one **ovarian follicle** culminates its maturation, whereas the remainder atrophy. After 14 days from the beginning of the cycle, the follicle is already mature and erupts on the surface of the ovary, ovulation. The oocyte, already converted into an ovum, breaks away from the ovary and enters the fallopian tube in search of spermatozoa to fertilize it. Due to the influence of the LH hormone, the remainder of the follicles are converted into corpus luteums or yellow bodies, which continue to secrete estrogen and also begin to produce progesterone. If there is no fertilization, the corpus luteum atrophies, is converted into a corpus albicans (white body), and ceases its hormonal production, which gives rise to menstruation. The cycle is repeated again and again until menopause begins as long as there is no pregnancy.

## Ovarian cycle

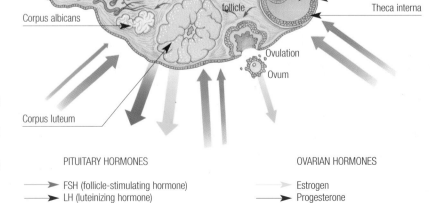

Stratum granulosum

Granulosa cells

Zona pellucida

Corona radiata

Liquor folliculi

Primary oocyte

Primary follicle

Atretic follicle

Graafian follicle

Theca externa

Theca interna

Corpus albicans

Ovulation

Ovum

Corpus luteum

PITUITARY HORMONES

→ FSH (follicle-stimulating hormone)
→ LH (luteinizing hormone)

OVARIAN HORMONES

→ Estrogen
→ Progesterone

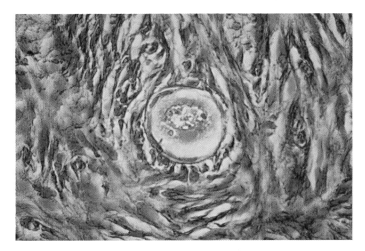

*Microscopic view of primitive ovarian follicles.*

## Phases of menstrual cycle

Infertile period

Ovulation

Fertile period

- Menstruation
- Lifetime of ovum
- Lifetime of spermatozoa
- Safe days

The menstrual cycle lasts on the average **28 days**, but it is absolutely normal for it to last between 21 and 35 days. During each menstrual cycle, the ovaries produce and release one mature ovum suitable for fertilization while, at the same time, they secrete female hormones that prepare the uterus for receiving the result of fertilization in case it should happen. They also have multiple effects on the entire female body. Since the expulsion of the ovum (ovulation) occurs toward the middle of the cycle and since both the lifetime of the ovum and that of the spermatozoa have limited duration, each menstrual cycle has a fertile phase when sexual relations can lead to pregnancy and an infertile period when fertilization is difficult.

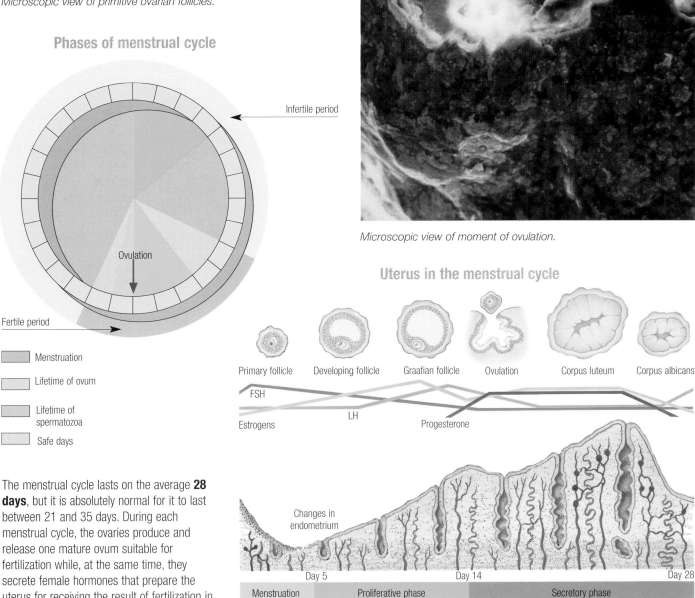

*Microscopic view of moment of ovulation.*

## Uterus in the menstrual cycle

Primary follicle    Developing follicle    Graafian follicle    Ovulation    Corpus luteum    Corpus albicans

FSH

Estrogens                LH                    Progesterone

Changes in endometrium

Day 5                    Day 14                        Day 28

| Menstruation | Proliferative phase | Secretory phase |

The function of the uterus is to receive the fertilized ovule and to house the fetus during pregnancy. For this purpose, it is prepared during each menstrual cycle under the influence of female hormones produced by the ovary. During the first part of the cycle, the estrogen that is secreted by the ovarian follicles gives rise to the proliferative phase. The mucous layer lines the interior of the uterus.

The endometrium gains thickness and prepares for the possible sheltering of a fertilized ovule.

After ovulation during the second part of the cycle, progesterone, processed by the luteal body, gives rise to the secretory phase. The endometrium continues to grow in size, its glands are activated, and its vascularization is developed notably.

# The breasts

**The breasts are paired and symmetrical organs situated in the** anterior part of the thorax that belong to the female reproductive system. Although they are secondary sexual features, they have an incredibly important function. They produce the milk that constitutes the basic food for the newborn.

During infancy, the breasts contain hardly any glandular tissue, whereas the nipples are only slightly pigmented and are practically flat. These characteristics persist in boys. In girls around the ages of eight or nine, the nipples begin to protrude, forming the so-called breast buds. This leads to the development of real breasts. They continue to develop due to the influence of estrogen throughout a period of some five to nine years and culminate in the transformation of the infantile breasts into adult breasts.

## Development of breasts

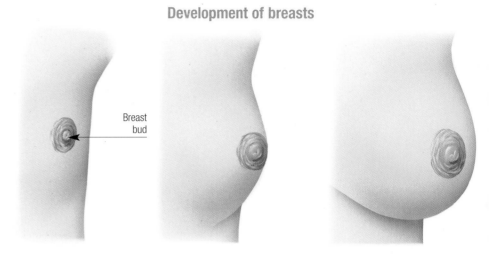

Breast bud

Ten years

Pubescent breast

Adult breast

## Internal structure of breast

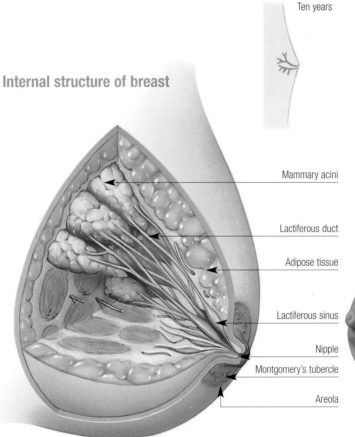

Mammary acini

Lactiferous duct

Adipose tissue

Lactiferous sinus

Nipple

Montgomery's tubercle

Areola

## Front view of breasts

Areola

Montgomery's tubercle

Nipple

Inframammary fold

Inside we find the **mammary gland**, which has the shape of a cluster and is made up of 15 to 20 lobes separated by accumulations of adipose tissue. The glandular tissue consists of units called acini, tiny sacs whose walls are made up of cells specialized in the secretion of milk. These acini empty into thin little ducts that converge upon each other and form others of larger caliber until they constitute thick ones, one for each lobe, called lactiferous ducts. These extend to the outside and, after dilating into the lactiferous sinuses, empty into the nipple. In the adult female, the breasts have a hemispherical or conical shape, although their shape, size, and appearance vary greatly. Approximately in

the center we find that the nipple, a circular, dark-colored prominence where the mammary glands empty, is surrounded by the areola, a circular zone of pigmented skin with likewise variable size and tone. The areola contains between 12 and 20 raised parts called Montgomery's tubercles, which correspond to some special sebaceous glands.

## Hormonal influences on the breast during the menstrual cycle

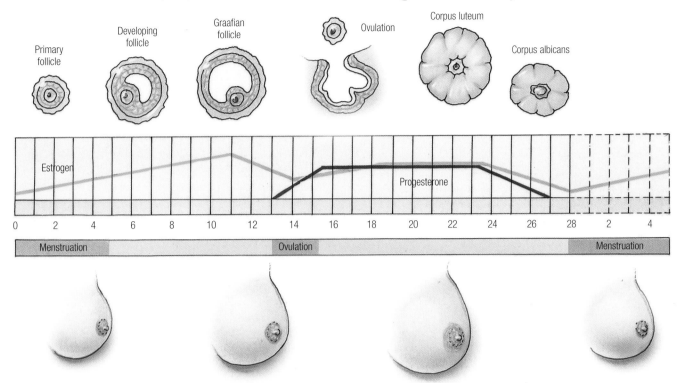

Under the influence of the ovarian hormones, the breasts undergo a series of changes during the menstrual cycle in preparation for a possible pregnancy. During the **first phase** of the cycle under the influence of the estrogen secreted by the developing ovarian follicles, there is a proliferation of the cells of the lactiferous ducts. A slight increase in the volume of the breasts as of the eighth day of the cycle can be noticed.

During the **second phase** of the cycle, after ovulation, the progesterone secreted by the corpus luteum induces the development of the glandular acini while, at the same time, it causes a certain retention of water in the body. This leads to a new increase in the volume of the breasts. Toward the end of the cycle, the breasts are enlarged and firmer. Coinciding with menstruation, the breasts lose size and firmness to start a new cycle all over again.

## Hormonal regulation of lactation

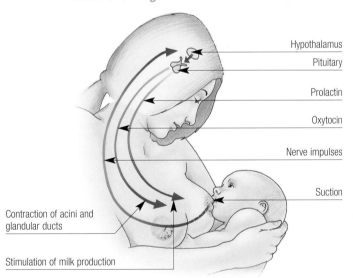

## Changes in the breast during pregnancy

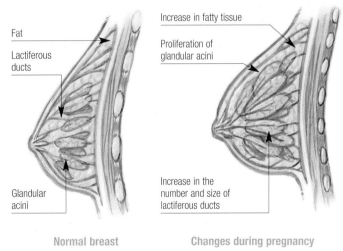

After delivery due to the influence of the **hormone prolactin**, the mammary glands are activated and begin to secrete milk. The stimulus coming from the suction of the baby causes the release of prolactin and maintains milk production throughout the entire time that the infant continues breast-feeding. In each breast, the baby's suction causes the release of **oxytocin**, a hormone that causes a contraction of the mammary glands and facilitates the delivery of milk through the nipple. When nursing is terminated and when the production of prolactin ceases, the mammary glands stop producing milk; they become involuted and return to the state of rest prior to pregnancy.

Throughout pregnancy, because of the response to the hormonal influences that are produced in the female's body under the influence of **progesterone**, the breasts undergo major changes. Starting with the first month of gestation, one can already detect a rather considerable proliferation of the lactiferous ducts, which leads to an increase in the volume of the breasts, one of the first signs of pregnancy. During the last phase of gestation, there is also a proliferation of the glandular acini. The breasts remain prepared to initiate their secretory activity.

# Sexual impulses and practices

biological and psychological as well as emotional aspects that are related to procreation and the continuity of the species as well as the giving and receiving of pleasure.

## Acquisition of sexual identity

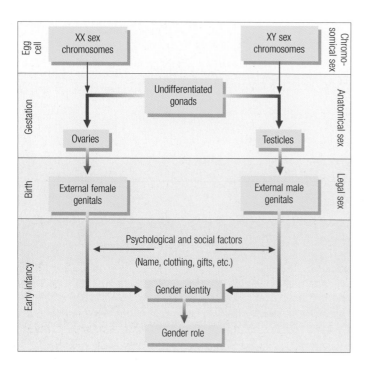

The human species has two sexes, male and female. From early infancy on, all individuals have a sense of belonging to one or the other gender. The sex of a person is established at the very moment of fertilization since it depends on the contribution of sex chromosomes from the germ cells. At the very instant of conception, therefore, the individual's chromosomal sex is determined. Due to hormonal influences, the genital organs, initially undifferentiated, develop in one way or another. Toward the third month of pregnancy, the individual's anatomical sex is established. The legal sex of the baby is established at the moment of birth, depending on the external genitals present. From that time, there will be many other conditioning and psychological factors until, toward the age of a year and half or two years, the gender identity is finally established. Although the female or male cannot yet distinguish anatomical differences with regard to genitals, the child will have a subjective sensation of belonging to one or the other sex. From then on, the individual will classify himself or herself as belonging to a particular sex. The individual follows the models prevailing in the environment, adopting the forms of conduct that are expected in view of his or her sex. This establishes the gender role, which is the public expression of sexual identity.

## Factors involved in the sexual impulse

Under normal conditions, all individuals have a certain degree of sexual impulse, an internal stimulus that acts as a source of fantasies that induces the individual, to a greater or lesser extent, to seek erotic situations and to be receptive to them. The sex impulse depends on the activity of the central nervous system and on hormonal factors. It appears that from the most primitive part of the cerebrum, the diencephalon, which is the seat of the hypothalamus and related to the limbic system, the spark that sets off sexual desire springs. The cerebral cortex, the most highly developed part of the brain, has the capacity to modulate sexual activity. The cerebral cortex is the site of intellectual functions. It generates internal stimuli and filters the external stimuli. It is therefore capable of boosting or canceling the sexual impulses.

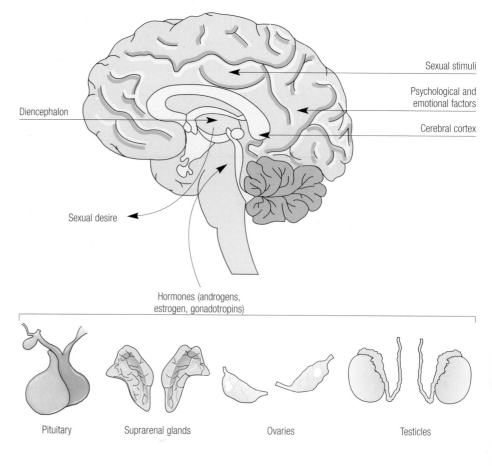

## Erogenous zones

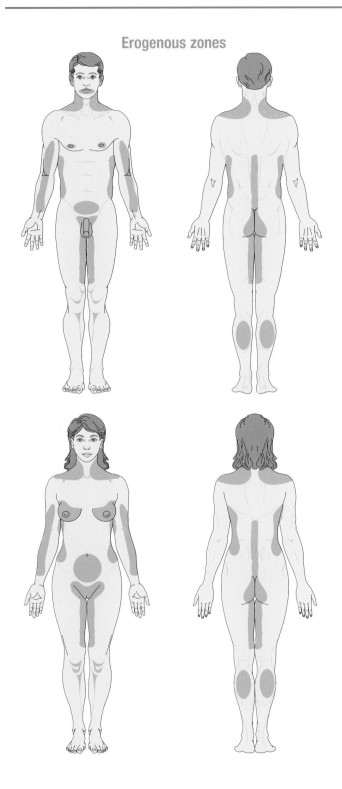

## Vaginal coitus

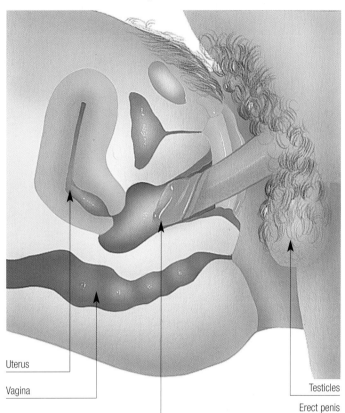

Uterus

Vagina

Testicles

Erect penis

## Sexual practices

In contrast to animals that have sexually stereotypical behavior, humans can enjoy very rich and varied sexuality. Common sexual practices include kissing and caressing of all kinds, stimulation of the genitals, both by hand and by mouth, which is now called oral sex, vaginal coitus, and anal coitus. This includes any practice pursued between freely consenting adults. The most widespread practices in most societies, however, although not the only ones, involves vaginal coitus, the introduction of the erect penis into the vagina. This is technically known as copulation and is usually referred to by the term sex act.

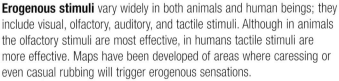

**Erogenous stimuli** vary widely in both animals and human beings; they include visual, olfactory, auditory, and tactile stimuli. Although in animals the olfactory stimuli are most effective, in humans tactile stimuli are more effective. Maps have been developed of areas where caressing or even casual rubbing will trigger erogenous sensations.

However, the capacity of the cerebrum to interpret stimuli as being erotic is very broad and varies to the point where it is completely impossible to list all of the factors that can trigger a sexual desire. Stimuli of a **psychological** nature are of utmost importance in humans. Although various purely physical stimuli can trigger a sexual response, the filter represented by the cerebrum very often turns out to be decisive in determining sexual interest. The greatest erogenous potential undoubtedly is found in the psychological area, especially in the **imagination**.

*Kisses on the mouth are one of the most widely used forms of sexual contact, both as a prelude to coitus and as a sexual activity itself. The photograph shows the sculpture by Auguste Rodin entitled* The Kiss.

# The human sexual response

After reaching a certain level of erotic stimulation that generates a certain degree of sexual tension, a series of sequential physical reactions will be automatically triggered. They constitute the so-called human sexual response, which is similar in all persons, both in males and in females, in spite of their obvious body differences.

Although it is difficult to outline the human sexual response since it is a succession of events that can take on different forms, one usually takes the work of the prestigious sexologists William Masters and Virginia Johnson, and, later on endorsed by other experts, as a basis for describing it. The human sexual response is arbitrarily divided into four phases: **excitation**, **plateau**, **orgasm**, and **resolution**.

## Nongenital reactions in the sexual response

Increase in respiration rate

Contraction of facial, abdominal, and intercostal musculature

Rise in blood pressure

Reddening of skin

Increase in heart rate

Increase in perspiration

In addition to the typical genital changes, a series of characteristic body reactions take place throughout the sexual response, both in men and in women.

## Male sexual response

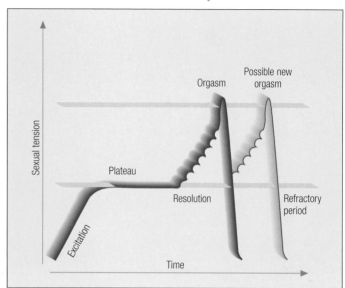

Sexual tension

Time

Orgasm

Possible new orgasm

Plateau

Resolution

Refractory period

Excitation

In men, the **excitation phase** is expressed in a neurovascular reflex that causes a great influx of blood to the genital zone and the swelling of the erectile bodies of the penis, causing erection. During the **plateau phase**, the state of erection of the penis continues with a degree of firmness that permits coitus. If sexual tension reaches a certain threshold, ejaculation takes place in a reflex fashion, accompanied by an intensive sensation of pleasure, the **orgasm**. Then during the **resolution phase**, the penis becomes flaccid as blood leaves it. After orgasm comes the refractory period during which a new ejaculation cannot take place even though a certain degree of erection persists.

## Changes in male genital organs during the sexual response

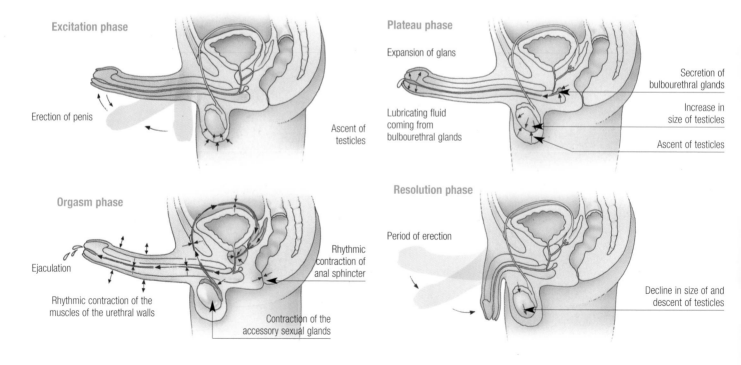

Excitation phase

Erection of penis

Ascent of testicles

Plateau phase

Expansion of glans

Lubricating fluid coming from bulbourethral glands

Secretion of bulbourethral glands

Increase in size of testicles

Ascent of testicles

Orgasm phase

Ejaculation

Rhythmic contraction of the muscles of the urethral walls

Rhythmic contraction of anal sphincter

Contraction of the accessory sexual glands

Resolution phase

Period of erection

Decline in size of and descent of testicles

## The female sexual response

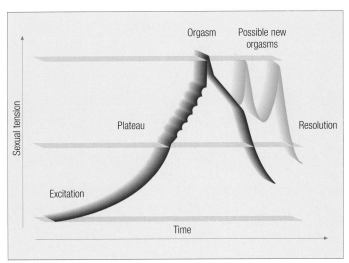

In females, the **excitation phase** features an increase in genital blood irrigation that is expressed in vasocongestion of the area and of the vaginal walls, a phenomenon that introduces a secretion of the fluid with the subsequent lubrication of the vagina. At the same time, the breasts are increased slightly in terms of volume while the nipples become erect. If sexual tension is continued, a stationary phase is reached during which there is an intensive vasocongestion of the outer third of the vaginal mucosa, the most sensitive zone, leading to the so-called orgasmic platform. The **orgasm** is triggered when sexual tension exceeds a certain threshold. That phase is a series of rhythmic contractions of the uterine musculature, of the muscles that surround the vagina, and of the anal sphincter, while at the same time, there is an intense sensation of pleasure. Then during the resolution phase, there is a progressive loss of blood in the genitals.

## Changes in genital organs and breasts during the female sexual response

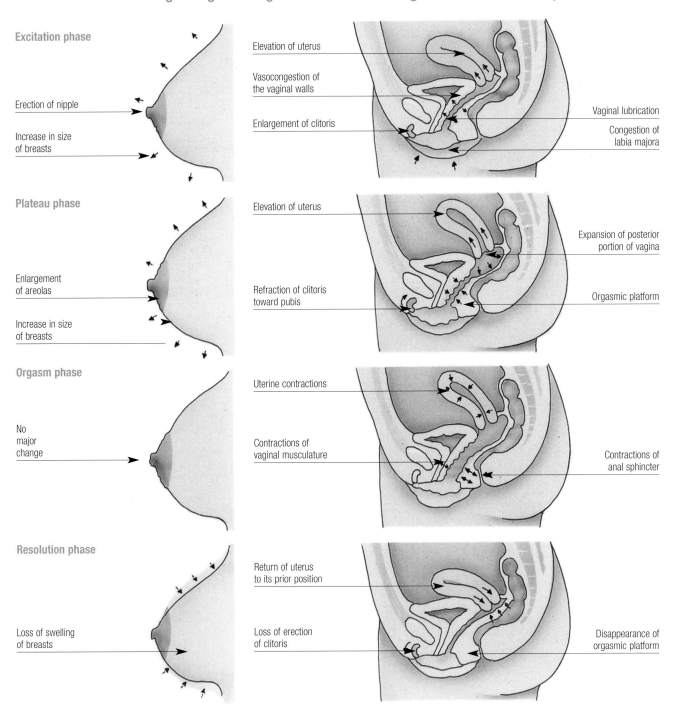

# Start of pregnancy

**Reproduction starts with the union of a maternal ovum and a** paternal spermatozoon. They merge into a zygote. The new human being will develop over a period of nine months in the uterus of the pregnant woman and will then enter the world and begin an independent life.

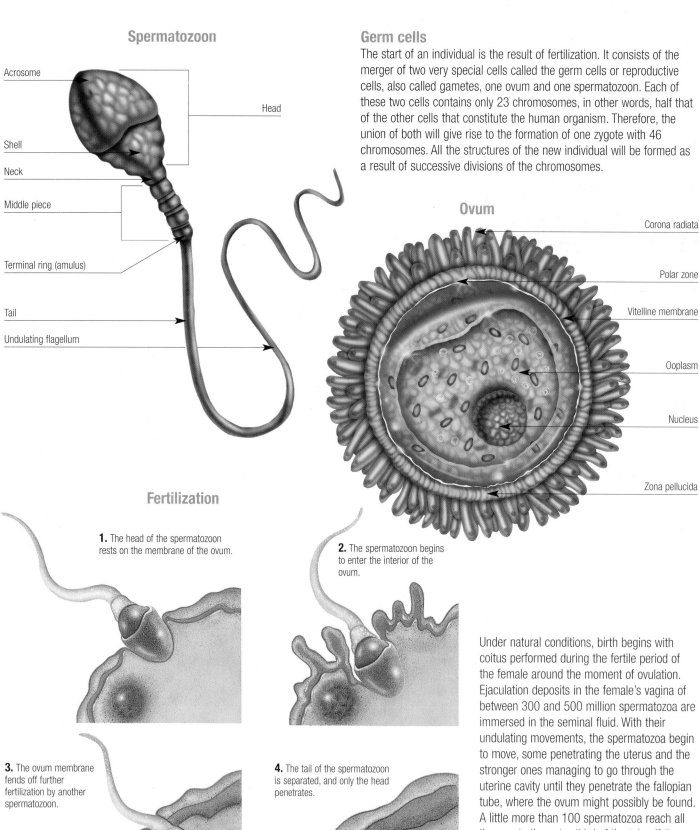

## Spermatozoon

Acrosome

Head

Shell

Neck

Middle piece

Terminal ring (amulus)

Tail

Undulating flagellum

## Germ cells

The start of an individual is the result of fertilization. It consists of the merger of two very special cells called the germ cells or reproductive cells, also called gametes, one ovum and one spermatozoon. Each of these two cells contains only 23 chromosomes, in other words, half that of the other cells that constitute the human organism. Therefore, the union of both will give rise to the formation of one zygote with 46 chromosomes. All the structures of the new individual will be formed as a result of successive divisions of the chromosomes.

## Ovum

Corona radiata

Polar zone

Vitelline membrane

Ooplasm

Nucleus

Zona pellucida

## Fertilization

**1.** The head of the spermatozoon rests on the membrane of the ovum.

**2.** The spermatozoon begins to enter the interior of the ovum.

**3.** The ovum membrane fends off further fertilization by another spermatozoon.

**4.** The tail of the spermatozoon is separated, and only the head penetrates.

Under natural conditions, birth begins with coitus performed during the fertile period of the female around the moment of ovulation. Ejaculation deposits in the female's vagina of between 300 and 500 million spermatozoa are immersed in the seminal fluid. With their undulating movements, the spermatozoa begin to move, some penetrating the uterus and the stronger ones managing to go through the uterine cavity until they penetrate the fallopian tube, where the ovum might possibly be found. A little more than 100 spermatozoa reach all the way to the outer third of the tube. If they bump into an ovum, they surround it and try to pass through the cell layers and membranes that are present in order finally to reach the ovum's surface and penetrate into its interior. This happens only once per ovum.

## Cleavage process

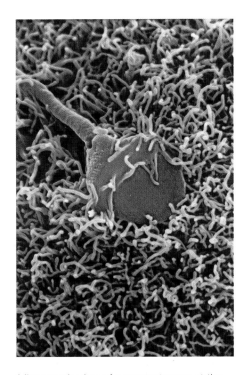

*Microscopic view of a spermatozoon at the moment it penetrates the interior of an ovum.*

Ovum

Spermatozoon

Fertilization

Egg cell

First cell division

Second cell division

Subsequent cellular divisions

Morula

After fertilization, the nuclei of the **ovum** and the **spermatozoon** merge, forming the zygote. A process called cleavage begins immediately. It is characterized by uninterrupted cell division. The two cells derived from the division of the zygote, the blastomeres, then divide in two, thus producing four cells. These then divide, and so on. After three days, the embryo already consists of 16 cells that form a conglomerate that looks somewhat like a tiny mulberry and is called a morula. Toward the fifth day, in addition to the cells dividing, the cells begin to differentiate and separate into layers. Fluid progressively

accumulates in the morula, displacing a group of cells toward one end. The embryo is now a blastula. It consists of two differentiated portions. The embryoblast is a group of cells from which the fetus will form. The trophoblast is a thin layer of cells that constitutes the boundary of the fluid-filled space, or blastocoel. The placenta will later form from the trophoblast.

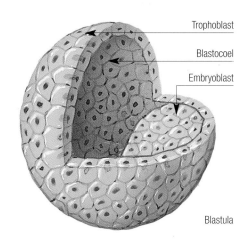

Trophoblast

Blastocoel

Embryoblast

Blastula

## Route of egg until implanted in uterus

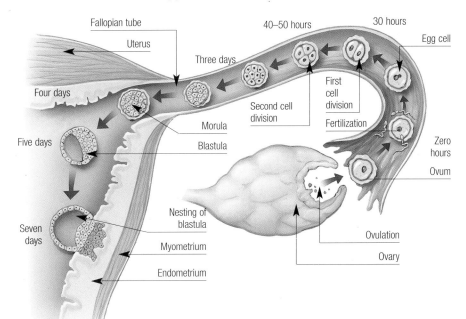

Fallopian tube

Uterus

40–50 hours

30 hours

Three days

Egg cell

Four days

First cell division

Second cell division

Five days

Morula

Fertilization

Blastula

Zero hours

Ovum

Nesting of blastula

Seven days

Ovulation

Myometrium

Ovary

Endometrium

During cleavage, the egg advances through the interior of the fallopian tube in the direction of the uterus. This occurs as a result of contractions in the musculature of the tube and the movements of tiny cilia of the cells in the mucosa that line it. The journey until the egg arrives inside the uterus cavity, which will shelter it for nine months, takes around four or five days. When the egg ends its travel through the fallopian tube, it arrives at the uterine cavity, where it remains free for two or three days while the endometrium finishes its preparations to receive it. Toward the seventh day after fertilization, the blastula rests gently on the surface of the endometrium in search of a site for settling down and obtaining nutritive substances. This is called nidation or implantation.

# Development of the embryo

**The first two months of pregnancy are the embryonic period** and represent a very important and extremely delicate phase of gestation. This is the stage when differentiation of the various tissues takes place and when practically all organs of the body are formed and begin to function.

## First phases of embryonic development

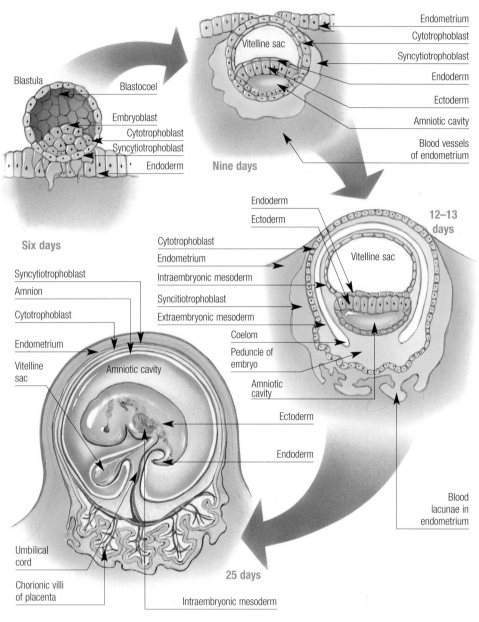

Blastula

Blastocoel

Embryoblast

Cytotrophoblast

Syncytiotrophoblast

Endoderm

**Six days**

Endometrium

Cytotrophoblast

Vitelline sac

Syncytiotrophoblast

Endoderm

Ectoderm

Amniotic cavity

Blood vessels of endometrium

**Nine days**

Syncytiotrophoblast

Amnion

Cytotrophoblast

Endometrium

Vitelline sac

Amniotic cavity

Endoderm

Ectoderm

Cytotrophoblast

Endometrium

Intraembryonic mesoderm

Syncitiotrophoblast

Extraembryonic mesoderm

Coelom

Peduncle of embryo

Amniotic cavity

Ectoderm

Endoderm

Vitelline sac

**12–13 days**

Blood lacunae in endometrium

Umbilical cord

Chorionic villi of placenta

**25 days**

Intraembryonic mesoderm

## Development of the embryo

Spectacular transformations follow in succession after the implantation of the blastula in the uterus. Within a few days, a simple conglomeration of cells is differentiated into diverse structures that will constitute the embryo and the membranes that will protect it and a placenta, the organ responsible for fetal oxygen supply and nutrition. The trophoblast will be differentiated into two layers, an outer layer (syncytiotrophoblast) and an internal layer (cytotrophoblast). A space is formed that is filled with fluid (amniotic cavity) whose size increases and is covered by a layer that will constitute the amnion or the amniotic sac (the water pouch). The sector corresponding to the incipient embryo (embryoblast) is converted into a disc with three cellular layers or blastodermic sheets: the ectoderm, which will give rise to the skin and the nervous tissue; the mesoderm, from which will derive the locomotor and circulatory systems; and the endoderm, which is the origin of the digestive, respiratory, and urinary systems.

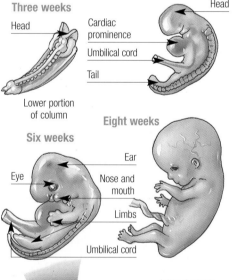

**Three weeks**

Head

Lower portion of column

**Four weeks**

Head

Cardiac prominence

Umbilical cord

Tail

**Six weeks**

Eye

**Eight weeks**

Ear

Nose and mouth

Limbs

Umbilical cord

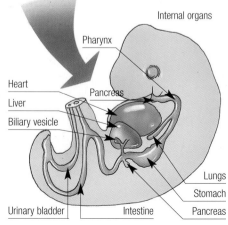

Internal organs

Pharynx

Heart

Liver

Biliary vesicle

Pancreas

Urinary bladder

Intestine

Lungs

Stomach

Pancreas

At the end of the first month, the embryo measures barely 5 mm and may not even weigh as much as 1 g, but it already has an elongated shape. At one end, it has a prominence corresponding to the head and some buds that will become the limbs. The nervous system as well as the circulatory system have already begun to take shape. The outlines of all of the other organs already appear during the second month, and the embryo's growth is stepped up. Length doubles during the fifth week, and the embryo takes on a shape that is reminiscent of a tadpole or a little seahorse with a head that is very big when compared with the body and a trunk that is slender. During this month, we can distinguish the head, the openings of the mouth and nose, and even the buds of the first teeth as well as the ear and eye buds. The limbs grow along the sides of the trunk, and the hands and feet develop. Next the organs of the digestive system, the liver, the pancreas, the kidneys, and the various muscles are formed. At the end of the eighth week, the embryo is 1–3 inches (3–5 cm) long and weighs between 2 and 3 g; it looks more human and has all of the organ systems.

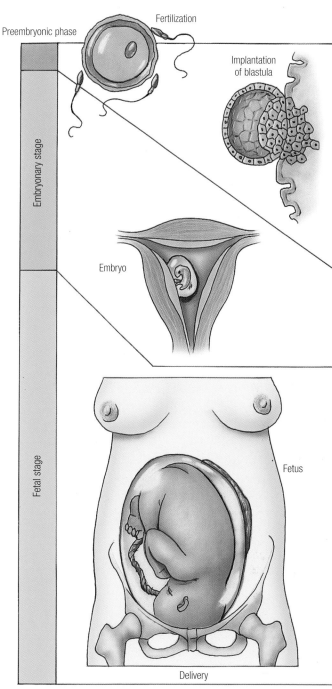

*Embryo, six weeks old.*

## Placenta

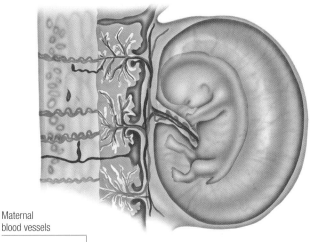

### Phases of intrauterine development

Preembryonic phase

Fertilization

Implantation of blastula

Embryonary stage

Embryo

Fetal stage

Fetus

Delivery

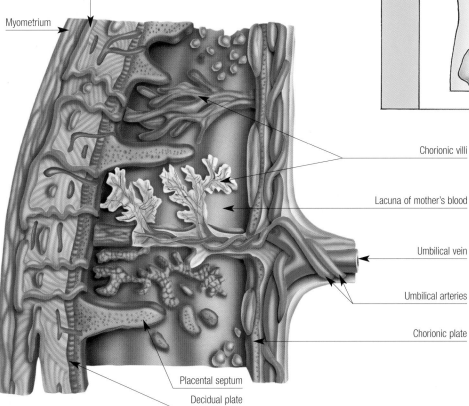

Maternal blood vessels

Myometrium

Chorionic villi

Lacuna of mother's blood

Umbilical vein

Umbilical arteries

Chorionic plate

Placental septum

Decidual plate

The placenta is an organ that exists only during gestation. Its task is to create a bridge between the maternal and fetal bodies. It is formed shortly after implantation from the external tissue of the embryo, called the chorion, and the uterine membrane that is adapted for pregnancy, the decidua basalis. After the placenta develops, the maternal vessels and the placental vessels will extend all the way to the fetus via the umbilical cord. An exchange takes place in the placenta between the mother's blood and the blood of the fetus, which are never in direct contact. Nutrients are passed from maternal circulation to fetal circulation, while metabolic wastes of the fetus move in the opposite direction and are then eliminated through the mother's body.

# Fetal Development

**The fetal stage, which comprises the major portion of gestation** from the third month of pregnancy until the moment of delivery, begins when all of the organs of the body are already sketched out or even formed. The organs only need to complete their maturation and begin functioning.

## Evolution of fetal development

*Growth in length and weight is constant and progressive throughout pregnancy.*

 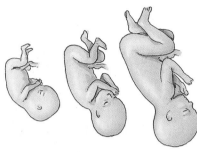 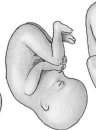 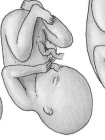 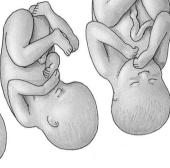

| Gestational stage | Second month | Third month | Fourth month | Fifth month | Sixth month | Seventh month | Eighth month | Ninth month |
|---|---|---|---|---|---|---|---|---|
| Length | 1–1.5 in. (3–4 cm) | 4 in. (10 cm) | 6 in. (16 cm) | 10 in. (25 cm) | 12.5 in. (32 cm) | 16 in. (40 cm) | 18.5 in. (47 cm) | 20 in. (50 cm) |
| Weight | 0.07–0.1 oz. (2–3 g) | 1 oz. (30 g) | 5 oz. (150 g) | 9–10 oz. (250–300 g) | 20 oz. (600 g) | 2.5–3 lb. (1.2–1.4 kg) | 4.5–5.5 lb. (2–2.5 kg) | 6.5–8 lb. (3–3.5 kg) |

## Fetal development in the maternal uterus

Throughout this entire period, the fetus undergoes a multitude of changes that follow each other uninterruptedly, although some characteristic transformations take place during every month. As it grows, the fetus adjusts to the space that is available in the uterine cavity. During the last month, the fetus settles into a head-down position, which is best for delivery.

 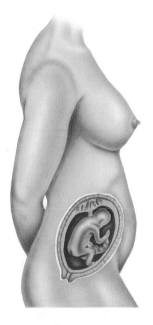 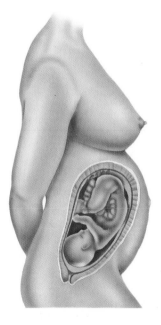 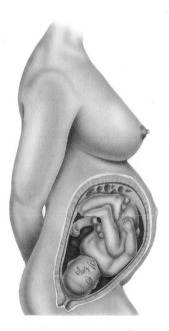

**THIRD MONTH**
The fetus is completely formed. Most of the organs are already functioning, and the period of very fast growth begins.

**FIFTH MONTH**
The fetus begins to move actively, already perceptible by the mother. The fetus reacts to stimuli.

**SEVENTH MONTH**
An important maturation of the internal organs takes place, and the fetus is practically capable of surviving if birth is imminent.

**NINTH MONTH**
The fetus is totally developed and fits into the maternal pelvis, preparing for the imminent delivery.

## Fetus in maternal uterus toward end of pregnancy

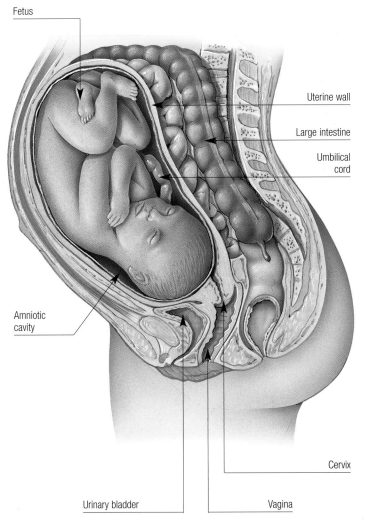

Fetus

Uterine wall

Large intestine

Umbilical cord

Amniotic cavity

Cervix

Urinary bladder

Vagina

Development of the bones is accelerated during the last weeks, and the skin becomes thicker. The head is much more in proportion, although it measures one-quarter of the body's total length. The ears are separated from the cranium, the nose is well-shaped, and the eyes take on a blue-gray color. The external genitals take on the definitive characteristics of the particular sex. In boys, the testicles descend from the abdomen and are situated in the scrotum; in girls, the vulva is almost covered by the major labia. Reflexes are very much perfected, especially, the suction reflex; soon it will be indispensable for correct nursing. The body of the fetus is already prepared for birth.

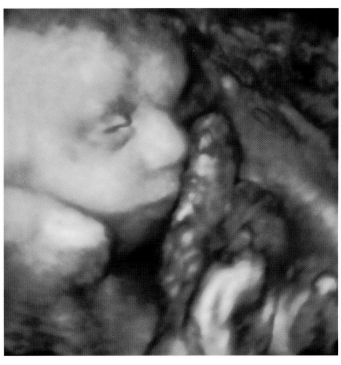

*Six-month-old fetus, floating in amniotic fluid.*

## Multiple births

Although in most pregnancies there is only one fetus, two or even more fetuses might form simultaneously in the mother's uterus. This gives rise to twins, triplets, and so on. In twins, two distinct ova can be fertilized by two different spermatozoa, causing the development of dizygotic, or fraternal, twins. Each one has a placenta of its own. They can be of the same sex or not and will have different appearances. At other times, the zygote derived from the merger of a single ovum and a single spermatozoon can divide into two or more fragments and as many embryos will form. Monozygotic, or identical twins, are then formed that share one single placenta and have the same genetic endowment. They are always of the same sex and look similar to each other.

### Twin pregnancy production mechanisms

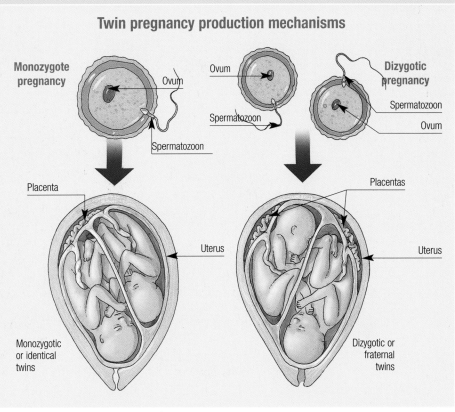

Monozygote pregnancy

Ovum

Spermatozoon

Ovum

Spermatozoon

Dizygotic pregnancy

Spermatozoon

Ovum

Placenta

Uterus

Placentas

Uterus

Monozygotic or identical twins

Dizygotic or fraternal twins

# Delivery

**Delivery is the culmination of pregnancy. After nine months of** waiting, powerful uterine contractions begin. They dilate the cervix and push the baby out of the uterus. The final portion of labor is when the placenta, or afterbirth, is delivered.

### How the head of the fetus fits into the maternal pelvis

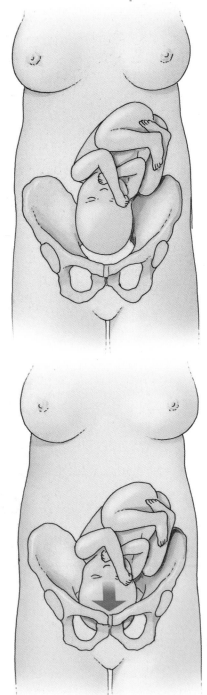

### Dilation and thinning of cervix

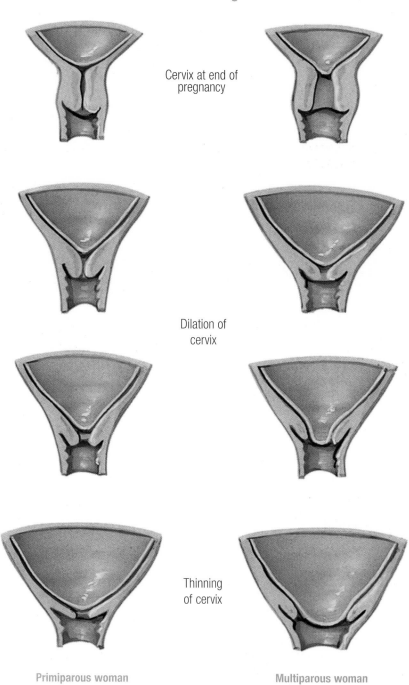

Cervix at end of pregnancy

Dilation of cervix

Thinning of cervix

Primiparous woman

Multiparous woman

For most of the pregnancy, the fetus floats freely inside the amniotic sac in the surrounding fluid. As the fetus grows, the available space shrinks and movements become restricted. As the moment of birth approaches, the fetus descends. The head should fit between the bones of the maternal pelvis. Everything is ready for the process of delivery to commence.

### Dilation phase

Labor begins with dilation and thinning of the cervix, creating a continuous channel with the vagina through which the baby can travel. This period is divided into two phases: latent and active. The latent phase, which is part of prelabor, extends from the start of regular contractions until the cervix has dilated to 3 cm. During this phase, the cervix also softens and thins. The active phase, which marks the start of active labor, begins once the cervix has dilated to 3 cm. The cervix continues to dilate until it has reached a diameter of 10 cm. Once it has dilated to 10 cm, the cervix is sufficiently wide to allow passage of the baby's head and shoulders.

## Delivery

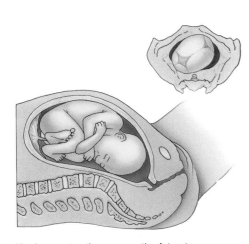

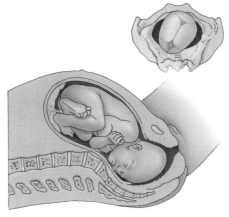

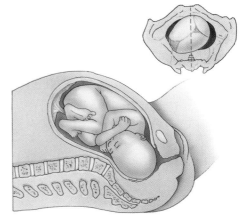

Uterine contractions cause the fetus to descend through the birth canal. The head is oriented so that it makes maximum use of the diameter of the maternal pelvis.

When the fetus begins to encounter resistance during the descent, the head bends so that the vertex of the cranium will be oriented toward the outside.

Subsequently, the fetus turns the head some 45º, orienting the forehead toward the maternal sacrum and the posterior part of the cranium toward the mother's pubic symphysis.

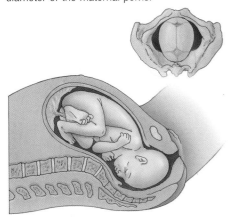

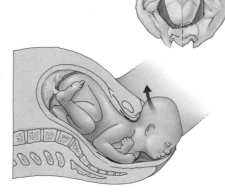

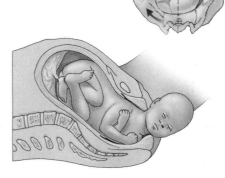

After rotation, the fetus extends the head using the maternal pubis as a point of support. At that instant, the head begins to become visible from the outside.

The fetus continues to extend the head with each uterine contraction, pushing toward the outside. First to exit is the brow, followed by the forehead and then the remainder of the face.

Once the head has come out, the fetus turns 90º to facilitate delivery of the trunk, with one shoulder upward and the other one downward.

During delivery, the baby must travel 4–5 inches (10–12 cm) through the birth canal to reach the maternal vulva. This occurs because of powerful uterine contractions. After dilating the cervix, the contractions push the baby through the cervix and vagina, which grow wider during the baby's descent.

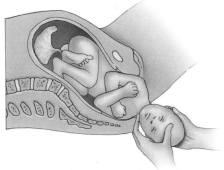

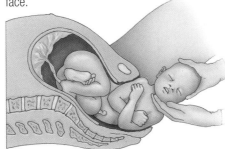

Then the shoulders emerge. The movement usually takes place spontaneously, but the doctor or midwife can assist by moving the head of the fetus downward to facilitate delivery of the shoulder oriented upward.

Then, while moving the head upward, the doctor or midwife assists in getting the lower shoulder out. Finally, the remainder of the body comes out quickly and without difficulty.

## Delivery of placenta

The last phase of labor is delivery of the placenta and associated membranes. As a matter of fact, uterine contractions continue and even increase in strength after the baby has been born. However, they are not as painful as those during active labor because the womb is already almost empty. These contractions lead to separation of the placenta from the uterine wall. After 5 to 30 minutes of contractions, the placenta is delivered.

# Hypothalamus and pituitary

**The endocrine system consists of a series of glands that, under** the direction of the hypothalamus and the pituitary, produce and release hormones into the blood. These substances act as chemical messengers to control metabolism, body growth, and development as well as the activity of various tissues and organs.

## Components of the endocrine system

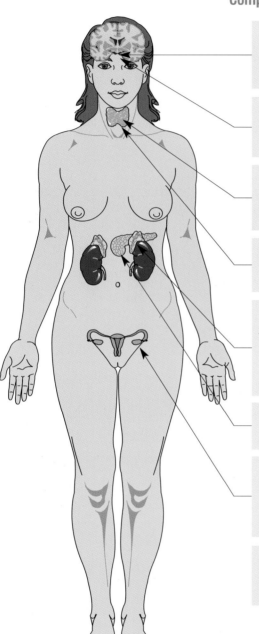

### Hypothalamus
It provides the link between the nervous system and the endocrine system, regulating the activity of the pituitary and, therefore, the entire endocrine system.

### Pituitary
Its hormones act directly on different tissues and control the activity of the other glands of the endocrine system.

### Thyroid
Its hormones stimulate the body's metabolic reactions that are essential to physical growth and mental development in children.

### Parathyroids
They produce hormones that participate in adjusting of calcium and phosphorus levels in the blood.

### Adrenals
They produce distinct hormones with diverse functions, some responsible for controlling the metabolism of nutrients or the hydrosaline balance of the body, while others act as mediators of the autonomic sympathetic nervous system.

### Pancreas
It secretes hormones that regulate glucose metabolism and blood concentration.

### Ovaries
Female gonads that produce estrogen and progesterone, hormones that are responsible for the development of secondary sexual characteristics and the activity of the female reproductive system.

### Testes
Male gonads producing testosterone, a hormone responsible for the development of secondary sexual characteristics.

## Location of hypothalamus and pituitary

The hypothalamus and the pituitary are two small structures situated in the base of the cerebrum. They have a particular anatomic relationship. On the one hand, neurons of the hypothalamus emit prolongations that extend to the posterior lobe of the pituitary (neurohypophysis). On the other hand, a network of venous vessels, or portal system, carries hormonal factors produced by the hypothalamus to the anterior lobe of the pituitary (adenohypophysis).

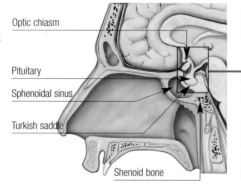

Optic chiasm

Pituitary

Sphenoidal sinus

Turkish saddle

Shenoid bone

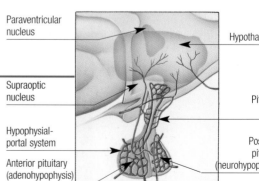

Paraventricular nucleus

Supraoptic nucleus

Hypophysial-portal system

Anterior pituitary (adenohypophysis)

Hypothalamus

Pituitary stalk

Posterior pituitary (neurohypophysis)

The hypothalamus has a wide variety of activities since it shelters the nerve centers that regulate thirst, appetite, body temperature, and sleep. This little gland, connected to various areas of the nervous system and therefore capable of receiving multiple stimuli, both physical and mental, stands out by virtue of another function: its role as a modulator of the endocrine system. It is the structure that really controls the activity of the internal glands and adapts their functions to the changing needs of the body.

## Functions of hypothalamus

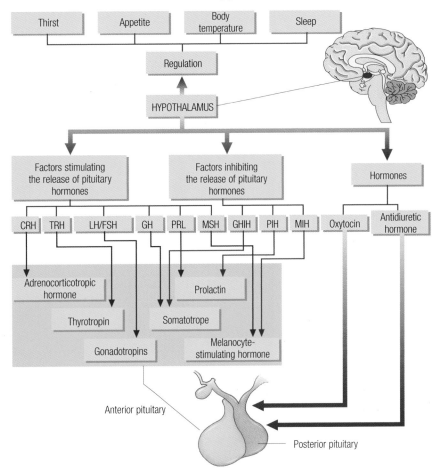

The pituitary regulates activity of the endocrine system through its hormones that act either on the tissues directly or on other endocrine glands. It produces seven hormones governing such fundamental matters as body growth and control of the activity of the thyroid, the adrenal cortex, and the gonads. In addition, it stores, and when necessary, releases two hormones produced by the hypothalamus, the antidiuretic hormone and oxytocin.

## Hormonal secretion of pituitary

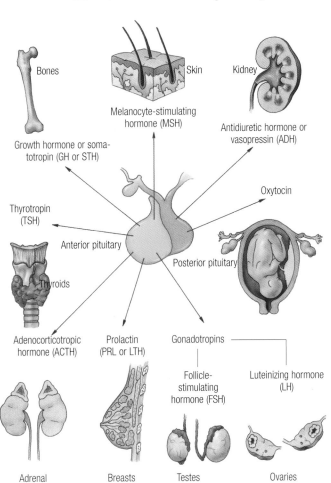

### Pituitary hormones

| Name | Abbreviation | Target organ | Function |
|---|---|---|---|
| Melanocyte-stimulating hormone | MSH | Skin | Stimulates the melanocytes that fabricate the pigment for skin color |
| Antidiuretic or vasopressin hormone | ADH | Kidney | Retains water in the kidneys; regulates arterial pressure |
| Growth hormone or somatotrope | STH or GH | Entire body | Stimulates the growth of the bones, muscles, and organs during infancy and puberty |
| Thyrotropin | TSH | Thyroid | Stimulates the activity of the thyroid gland |
| Oxytocin | | Uterus | Causes the contractions of the uterus during delivery |
| Adrenocorticotropic hormone | ACTH | Adrenal glands | Stimulates production of corticosteroids in the adrenal glands |
| Prolactin | PRL | Breasts | Causes secretion of milk after delivery |
| Gonadotropins<br>• follicle-stimulating hormone<br>• luteinizing hormone | FSH<br><br>LH | Gonads (ovaries and testes) | Regulates the maturation of the spermatozoa and the ova as well as the production of sex hormones |

# Thyroid and parathyroid glands

**The thyroid gland is a small endocrine gland situated in the** anterior part of the neck whose secretions regulate metabolism. The parathyroids, four tiny glands, are so called because they are situated in the posterior face of the thyroid, producing a hormone that participates in the regulation of blood calcium levels.

## Anatomy of the thyroid gland

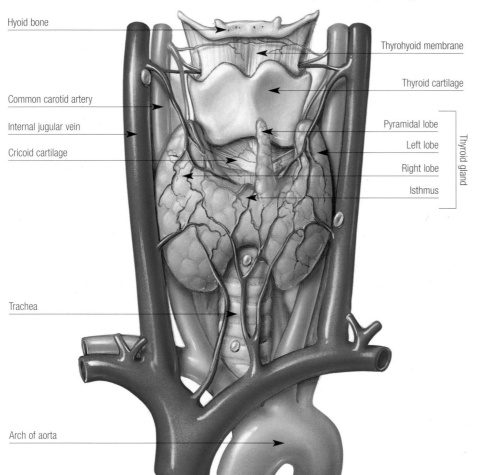

Hyoid bone

Common carotid artery

Internal jugular vein

Cricoid cartilage

Trachea

Arch of aorta

Thyrohyoid membrane

Thyroid cartilage

Pyramidal lobe

Left lobe

Right lobe

Isthmus

Thyroid gland

Made up of two lateral lobes that surround the beginning of the trachea, the thyroid gland is linked by a narrow portion of tissue called the isthmus. Sometimes there is also a small superior prolongation called the pyramidal lobule.

## Function of thyroid gland

The thyroid gland produces hormones that stimulate cellular functions and therefore activate metabolism and the production of heat. During infancy, the thyroid hormones decisively influence maturation of the nervous system and body growth, affecting physical and mental development. The two principal thyroid hormones, which contain iodine, are thyroxine ($T_4$) and triiodothryronine ($T_3$). These hormones perform a similar action. In practically all tissues, they cause an increase in metabolic reactions. The thyroid gland also produces another hormone, calcitonin, which is involved in the regulation of calcium blood levels.

## Production of thyroid hormones

Due to the stimulus from thyrotropin, the cells of the thyroid gland take iodine (I) from the blood and synthesize a protein called thyroglobulin. Within the cells, iodine is coupled to the molecules of thyroglobulin and gives rise to the formation of two products: monoiodotyrosine, which has one atom of iodine, and diiodotyrosine, which has two. A subsequent coupling of these two products gives rise to the formation of $T_3$, which consists of three atoms of iodine, and $T_4$, which has four atoms of iodine. Once processed, hormones are stored in the thyroid gland until they are released to the blood circulation when needed.

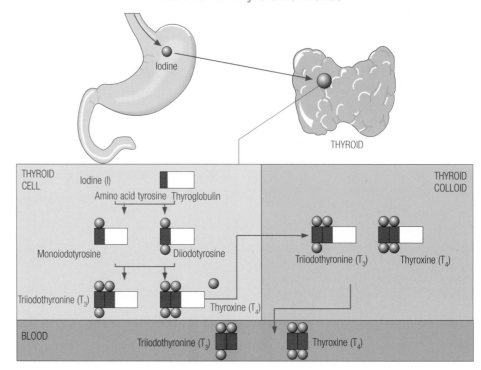

## Regulation of thyroid activity

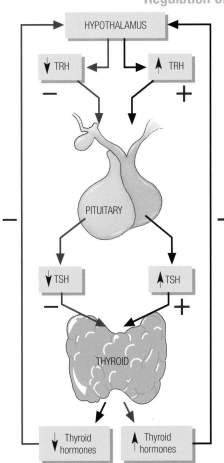

Thyroid activity is governed by the hypothalamic-pituitary axis. The gland responds to the stimulus from the hormone thyrotropin (TSH) produced by the pituitary. The production of TSH, in turn, depends on the thyrotropin-releasing factor (TRH) made by the hypothalamus. The production of thyroid hormones is based on a mechanism of negative feedback. The blood concentration of a hormone is the principal factor that affects the activity of the hypothalamus and the pituitary.

When thyroid hormone blood levels are high, the hypothalamus detects this situation and secretes less TRH and stops stimulating the pituitary to produce TSH. The production of thyrotropin then drops; the thyroid gland is less stimulated and reduces its hormonal production. When the thyroid hormone levels drop excessively, the hypothalamus increases its secretion of TRH, which acts upon the pituitary and gives rise to an increase of TSH. The production of thyrotropin thus goes up, and that stimulates thyroid activity.

## Location of parathyroid glands

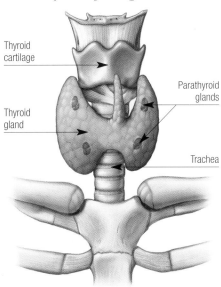

Parathyroid glands are four small, round or oval, yellowish formations, with a diameter of a few millimeters and a total weight of 25–40 mg. Individually, these are the smallest organs of the entire body. They are in pairs on either side of the trachea. Each thyroid lobe contains two parathyroid glands: one in the upper part, which is farther to the outside, and another in the lower part, farther inside.

## Function of parathyroid glands

The parathyroid glands produce parathyroid hormone, or parathormone (PTH). Together with calcitonin produced by the thyroid gland and vitamin D, PTH is involved in the adjustment of calcium blood levels. The parathyroid hormone tends to increase calcium blood levels, acting on the bones, the kidneys, and the digestive tract. In the bones, it stimulates the activity of the osteoclasts and, therefore, promotes the destruction of bone tissue. As a result, the bones release some of the stored calcium into the blood. In the kidneys, PTH promotes the reabsorption of calcium, which reduces the urinary elimination of this mineral and raises levels in the blood. Through the activation of vitamin D in the kidneys, it promotes intestinal absorption of calcium contained in food in the digestive tract.

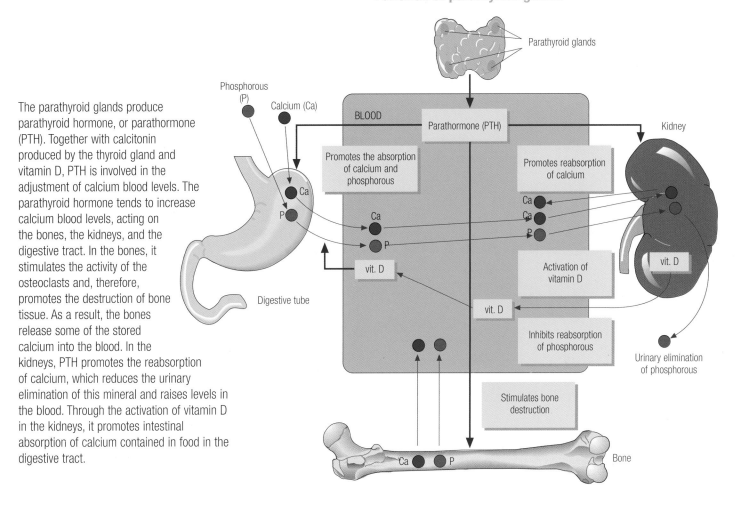

# Adrenal glands and pancreas

**The adrenal glands process hormones performing actions as** diverse as the adjustment of arterial pressure, balance of fluids and minerals, metabolism of nutrients, and development of secondary sexual characteristics. The pancreas produces two hormones, insulin and glucagon, that regulate the blood glucose level.

## Location of adrenal glands

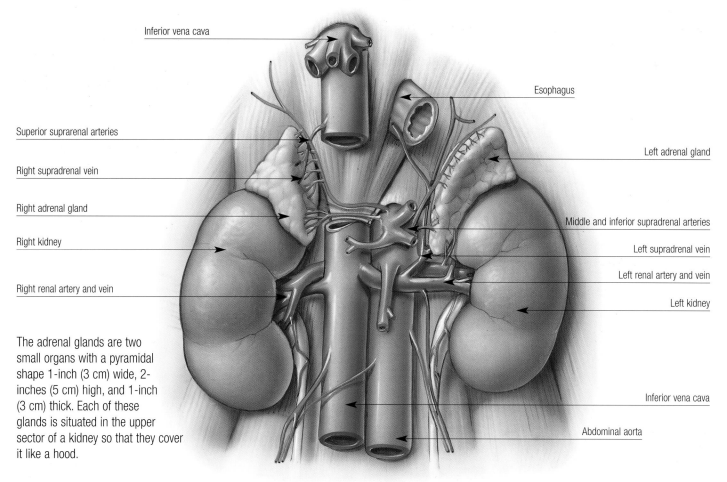

Inferior vena cava

Superior suprarenal arteries

Right supradrenal vein

Right adrenal gland

Right kidney

Right renal artery and vein

Esophagus

Left adrenal gland

Middle and inferior supradrenal arteries

Left supradrenal vein

Left renal artery and vein

Left kidney

Inferior vena cava

Abdominal aorta

The adrenal glands are two small organs with a pyramidal shape 1-inch (3 cm) wide, 2-inches (5 cm) high, and 1-inch (3 cm) thick. Each of these glands is situated in the upper sector of a kidney so that they cover it like a hood.

## Cross section of an adrenal gland

Each adrenal gland is enveloped by a capsule of connective tissue surrounded by adipose tissue. Two completely distinct parts can be differentiated: the cortex and the medulla. The **adrenal cortex**, situated immediately below the capsule, is the major part of the gland. A thick layer of tissue, it is made up of epitheloid cells, which fabricate corticosteroid hormones. The cortex consists of three differentiated layers of tissue. The outermost is the glomerular area; the intermediate one is the fascicular area; the innermost zone is the reticular area. Each of these areas of the adrenal cortex produces different hormones with definitely distinct functions. The **adrenal medulla**, which occupies the central region of the gland, is made up of nerve tissue and consists of cells that specialize in the production of hormones whose secretion is

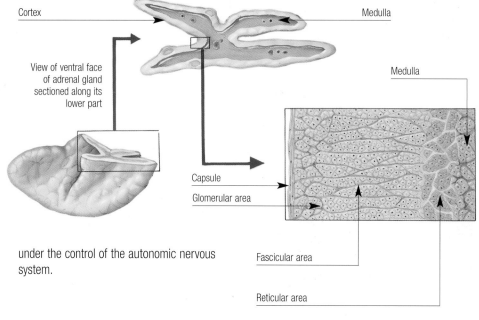

Cortex

Medulla

View of ventral face of adrenal gland sectioned along its lower part

Medulla

Capsule

Glomerular area

Fascicular area

Reticular area

under the control of the autonomic nervous system.

## Activity of suprarenal cortex

Stress

CRH

Hypothalamus

ACTH

Pituitary

Androgens

Muscle

Formation of proteins

Destruction of proteins

Cortisol

Liver

Formation of glucose

Adipose tissue

Formation of fat

White blood cells

Anti-inflammatory action

Adrenal gland

Aldosterone

Kidney

Retention of sodium, elimination of water

Retention of sodium and water

Stomach

Gastric secretion

Blood

The adrenal cortex produces various corticosteroid hormones. One group of these hormones is the **mineralocorticoids**, whose principal component is aldosterone. It participates in the regulation of fluid and salt balance (sodium and potassium). Aldosterone works in the kidney and adapts the loss of water and salts through urine to the needs of the body.

Another group is made up of the **glucocorticoids**. Its principal component is the hormone cortisol or hydrocortisone, which regulates metabolism of energy nutrients, exerts powerful anti-inflammatory action, and has immunosuppressant effects.

A third group consists of the **androgens**, hormones that promote the development of male sexual characteristics and the growth of muscle tissue.

## Endocrine pancreas

In addition to producing a secretion rich in enzymes discharged into the small intestine and performing a fundamental role in the digestive process, the pancreas also acts as an endocrine gland. It produces two hormones involved in the metabolism of carbohydrates and regulating the glucose levels in the blood, in other words, glycemia. This twin activity is reflected in the organ's anatomy. The endocrine pancreas is made up of a series of microscopic groups of cells, the islets of Langerhans, distributed throughout the organ and primarily in the tail, surrounded by an accumulation of pancreatic tissue that produces the digestive secretions, the pancreatic acini. The islets of Langerhans have two types of special cells: the alpha cells, which produce glucagon, and the beta cells, which produce insulin.

## Actions of insulin and glucagon

Glucagon   Alpha cell   Pancreas   Beta cell   Insulin

Blood

Glucagon

Glucose

Glucose

Glucose

Glucose

Glucose

Insulin

Glucose

receptor

Glucose

Fatty acids

Adipose cell

receptor

Glucose — Glycogen

Glucose   Glucose

Amino acids — Proteins

Glycogen

Fatty acids

Liver cell

Muscle cell

## Islet of Langerhans

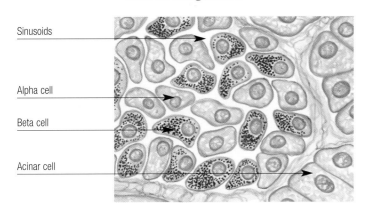

Sinusoids

Alpha cell

Beta cell

Acinar cell

Insulin and glucagon have antagonistic effects. Insulin promotes the entry of glucose circulating in the blood to the interior of the body's cells, which use these nutrients as the main energy source. Thus, the action of insulin has a hypoglycemic effect that causes a drop in the blood glucose concentration. It also acts on liver cells and muscle cells so that they may transform the glucose into glycogen, the organism's energy reserve, promoting the conversion of glucose into fatty acids and lipids, which accumulate in the adipocytes, and promoting protein synthesis. Glucagon, on the other hand, has a hyperglycemic effect since it promotes the degradation of glycogen stored in the liver cells and the passage of glucose into the blood.

# Lymphatic organs and immunity

**The immune system consists of a group of lymphatic tissues and** organs where the white blood cells are produced and mature. White blood cells and the antibodies they produce defend the body against potentially harmful foreign elements, especially germs, from the external environment.

## Lymphatic organs

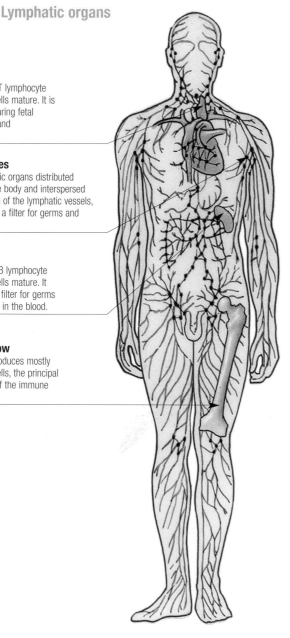

**Thymus**
In this organ, T lymphocyte white blood cells mature. It is most active during fetal development and infancy.

**Lymph nodes**
Small lymphatic organs distributed throughout the body and interspersed along the path of the lymphatic vessels, functioning as a filter for germs and impurities.

**Spleen**
In this organ, B lymphocyte white blood cells mature. It also acts as a filter for germs and impurities in the blood.

**Bone marrow**
Tissue that produces mostly white blood cells, the principal components of the immune system.

## The thymus

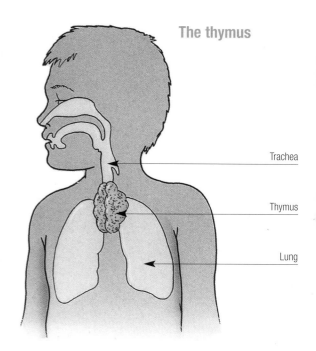

Trachea

Thymus

Lung

The thymus is a small organ situated in the center of the chest behind the sternum. Its function and development are unique. It becomes active during gestation and subsequently atrophies. Beginning in infancy and ending with puberty, lymphocytes formed in the bone marrow move to the thymus and mature there. Now these T lymphocytes are capable of taking part in the immune reaction. The thymus is active and increases in size while the body is still immature. Its activity ends during puberty, around 14–16 years of age, when maturation of the immune system is complete. The thymus attains a maximum weight of 1.5 ounces (45 g). After puberty, the organ progressively atrophies until it weighs only about 0.5 ounce (15 g) in an adult.

The body contains numerous lymph nodes distributed throughout the body. They are of the utmost importance in defense since they shelter a large quantity of white blood cells that are responsible for detecting and neutralizing or destroying germs or impurities transported by the lymph vessels that drain the body's tissues. Each node is made up of a capsule of connective tissue from which extend trabeculae that divide the node into various portions where there are sinuses full of white blood cells. Afferent lymph vessels arrive at the node, transporting the lymph collected in the tissues, which is then filtered inside and relieved of any harmful or potentially dangerous elements in order to exit through some efferent lymph vessels and continue on their way to the circulatory system. Since the lymph nodes are situated in strategic places throughout the body, their action prevents the spread of harmful agents throughout the entire organism.

## Lymph node

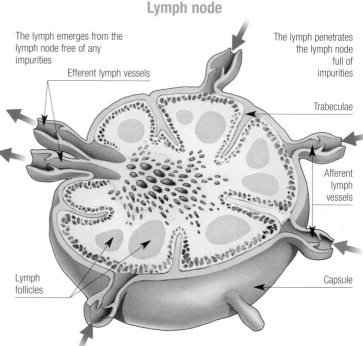

The lymph emerges from the lymph node free of any impurities

Efferent lymph vessels

The lymph penetrates the lymph node full of impurities

Trabeculae

Afferent lymph vessels

Lymph follicles

Capsule

## Nonspecific immune response

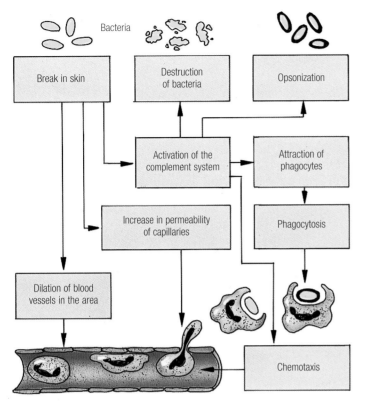

The body has a series of resources to protect itself in a nonspecific manner against attack by potentially pathogenic germs. First of all, there are protective barriers. These include the skin, which prevents the entry of these agents into the body. There are also the various fluids containing enzymes capable of destroying many microbes, such as nasal mucus, saliva, and tears. If the germs overcome these lines of defense, they confront the action of the phagocytes, white blood cells that run throughout the body and ingest and digest any foreign particle they encounter. Germs also confront the entire group of complement proteins capable of attacking the walls of the germs and destroying them or promoting the action of the white blood cells.

## Specific immune response

When the body is overcome by the first defense mechanisms, it triggers an immune reaction defending itself exclusively against each aggressor agent. This defensive response, which is produced by the white cells, is based on recognizing structural elements of the foreign agent called antigens, activating a series of cellular and humoral mechanisms to destroy or neutralize the aggressor. Cellular immune response is taken care of by the T-lymphocytes of different varieties. One detects the germ and secretes chemical substances that generate an alarm signal in the area, while others act as "assassin cells" and attack and destroy the microbe. The humoral immune response is provided by the B-lymphocytes which, in response to the alarm signal, multiply and are converted into plasmatic cells responsible for processing antibodies. These are gammaglobulins that are coupled to the antigens of the attacking germ and facilitate the attack of immunity cells present in the area.

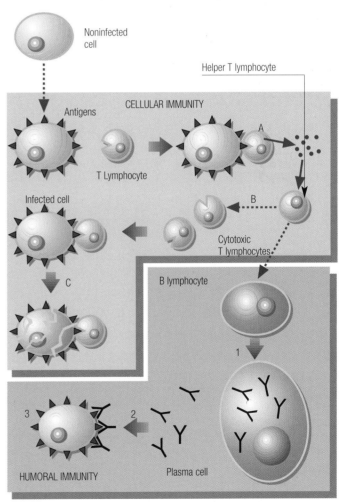

**A.** The T lymphocyte recognizes the antigen and releases chemical substances.
**B.** The helper T lymphocytes activate the cytotoxic T lymphocytes and the B lymphocytes.
**C.** The cytotoxic T lymphocytes bind with the antigens and destroy the cell.
**1.** The B lymphocytes are differentiated into plasma cells.
**2.** The plasma cells release antibodies.
**3.** Antibodies bind with the antigens and destroy or inactivate the bacteria.

## Allergy

**A**n allergy is an alteration of the immune system that reacts to contact substances that are harmless to the majority of the population. These substances trigger an inflammatory response, such as sneezing, difficulty breathing, and skin eruptions. Allergy-producing substances range from various pollens to foods, such as nuts, and medications.